Benjamin Huh's

SAT® SUBJECT TEST

MATH LEVEL 2

Benjamin Huh, M.Ed.

University of California - Berkeley / Applied Math, BA

University of Washington / Math Education, M Ed.

Director / Berkeley Prep & Consulting, Seattle, WA

About the Author

The Author, Benjamin Huh has B.A. degree in Applied Mathematics from UC Berkeley and M.Ed degree in Mathematics Education from University of Washington. He has been teaching secondary and college mathematics for more than 20 years, including algebra I, algebra II, geometry, pre-calculus, AP calculus AB & BC, AP Statistics as well as advanced math in multi variable calculus, differential equation and linear algebra. He has been director of Berkeley Learning Center for more than decades and taught many High school students in the U.S. and Korea helping them not only in the subject of Mathematics producing many perfect scored students in SAT I and II math, but also in consulting and counselling of the US College Admission and Application for his students to gain admissions from the highly selective colleges and universities in the nation including IVY League schools and top tier schools.

Benjamin Huh

University of Washington, M.Ed.
University of California - Berkeley, Mathematics BA

- 2015-Present   Berkeley Prep & Consulting - Bellevue, WA, Director/Math Instructor
- 2009-2014   Berkeley Learning Center - Irvine, CA, Director/Math Instructor
- 2004-2008   Berkeley Learning Center - Gangnam/Seoul, Korea, Director/Math Instructor
- 1999-2004   Berkeley Learning Center - Seattle/Tacoma, WA, Director/Math Instructor

Acknowledgments

I thank my friend, Andrew An, for his work on editing, formatting
and the preparation of the manuscript of this book.

ISBN-13: 978-0996564700
ISBN-10: 0996564705

Contents

PART 2
MODEL TESTS

1. Math Level 2 Overview of the Contents*

50 multiple-choice questions and topics covered- as follow;

Content	Approximate Number of Questions
Number and Operations Operations, ratio and proportion, complex numbers, counting, elementary number theory, matrices, sequences, series, vectors	5-7
Algebra and Functions Expressions, equations, inequalities, representation and modeling, properties of functions (linear, polynomial, rational, exponential, logarithmic, trigonometric, inverse trigonometric, periodic, piecewise, recursive, parametric)	22-24
Geometry and Measurement	15-18
Coordinate Lines, parabolas, circles, ellipses, hyperbolas, symmetry, transformations, polar coordinates	3-4
Three-Dimensional Solids, surface area and volume (cylinders, cones, pyramids, spheres, prisms), coordinates in three dimensions	3-4
Trigonometry Right triangles, identities, radian measure, law of cosines, law of sines, equations, double angle formulas	9-10
Data analysis, Statistics and Probability Mean, median, mode, range, interquartile range, standard deviation, graphs and plots, least squares regression (linear, quadratic, exponential), probability	4-5

2. Difficulty Level of Mathematics Level 2*

- More than three years of college-preparatory mathematics, including two years of algebra, one year of geometry, and elementary functions (pre-calculus) and/or trigonometry.
- If you have had preparation in trigonometry and elementary functions and have attained grades of B or better in these courses, then taking Level 2 is appropriate.

3. Types of Calculators Recommended*

- Bring a calculator that you are used to using. It may be a scientific or a graphing calculator.
- We recommend the use of a graphing calculator over a scientific calculator because a graphing calculator may provide an advantage on some questions. Do NOT Bring These Unacceptable Calculators.
- Models that have wireless, Bluetooth, cellular, audio/video recording and playing, camera, or any other smart phone type feature.
- Models that can access the Internet.
- Models that have QWERTY, pen-input, stylus**, or touch-screen capability; require electrical outlets; or use paper tape (e.g., TI-92 Plus, Voyage 200, Palm, PDAs, Casio Class Pad)
- Models that "talk" or make unusual noises.

    (** The use of the stylus with the Sharp EL-9600 calculator is not permitted.)

1

4. Calculator Use and Strategy*

You do not need to use a calculator to solve every question, and it is important to know when and how to use one. First decide how you will solve a problem; then determine whether the calculator is needed.

- You'll need a calculator for 40 to 50 percent of the questions on Level 1 and for 55 to 65 percent of the questions on Level 2.

- For the rest of the questions, there is no advantage, perhaps even a disadvantage, to using a calculator.

- Do not round any intermediate calculations. When you get a result from the calculator for the first step of a solution, keep the result in the calculator and use it for the second step. If you round the result from the first step, your answer may not be one of the choices.

- You may not use a calculator for other Subject Tests and must put it away when not taking a mathematics test.

- Bring a calculator that you are used to using. It may be a scientific or a graphing calculator, but if you're comfortable with both, bring a graphing calculator. The most important consideration is your comfort level with the calculator. Test day is not the time to start learning how to use a new calculator, even if it has more capabilities.

- Verify that your calculator is in *good working condition* before you take the test. You may bring batteries and a backup calculator to the test center. Remember, no substitute calculators or batteries will be available at the test center. You can't share calculators with other test takers.

- If you are taking the Mathematics Level 1 test, make sure your *calculator is in degree mode ahead of time* so you won't have to worry about it during the test.

- If your calculator malfunctions at the test center, and you don't have a backup calculator, you must tell your test supervisor when the malfunction occurs. You can choose to cancel your scores on the test.

- If you are using a calculator with large characters (one inch high or more) or a calculator with a raised display that might be visible to other test takers, you will be seated at the discretion of the test supervisor.

- You may *not* use your calculator for sharing or exchanging, or removing part of a test book or any notes relating to the test from the test room. Such action *may be grounds for dismissal, cancellation of scores or both.* You do not have to clear your calculator's memory before or after taking the test.

(* Based on information from The College Board)

SAT II Math – Formula and Notes

1. Quadratic Function,

Vertex Form:

$y = ax^2 + bx + c = a(x - h)^2 + k$

Vertex (h, k), where $x = \dfrac{-b}{2a}$, or called a line of symmetry, and $V = (x, f(x))$

(eg. 1) What is the vertex of $y = f(x) = x^2 - 3x + 2$?

$y = f(x) = x^2 - 3x + 2$ has vertex at $x = \dfrac{-b}{2a}$.

$\therefore x = \dfrac{-(-3)}{2} = 1.5$

$\therefore f(1.5) = (1.5)^2 - 3(1.5) + 2 = 2.25 - 4.5 + 2 = -0.25$

$\therefore$ vertex, $v = (1.5, -0.25)$

Quadratic Equation:

Given $ax^2 + bx + c = 0$, $x = \dfrac{-b \pm \sqrt{b^2 - 4ac}}{2a}$.

And suppose its zeros are x_1, and x_2, then, the sum of $x_1 + x_2 = \dfrac{-b}{a}$ and the product $x_1 \cdot x_2 = \dfrac{c}{a}$.

(eg. 1) In the equation $x^2 + kx + 48 = 0$, one root is twice the other root. The value(s) of k is (are)?

To solve this problem, we need to refer to the following formula from the quadratic eq.:
Given $ax^2 + bx + c = 0$, suppose its zeros are x_1, and x_2.

Then, the sum of $x_1 + x_2 = \dfrac{-b}{a}$, and the product $x_1 \times x_2 = \dfrac{c}{a}$.

Now, from our equation, $x^2 + kx + 48 = 0$, we get $a = 1$, $b = k$ and $c = 48$.

Also, the problem states that one root is twice of the other root, this, $x_2 = 2x_1$.

Now, the sum $x_1 + x_2 = x_1 + (2x_1) = 3x_1 = \dfrac{-b}{a}$, or $3x_1 = \dfrac{-k}{1} = (-)k$, or $x_1 = \dfrac{-k}{3}$.   --- eq.(1)

Also, the product $x_1 \times x_2 = x_1 \times (2x_1) = 2x_1^2 = \dfrac{c}{a}$, or $2x_1^2 = \dfrac{48}{1} = 48$, or $x_1^2 = 24$.   --- eq.(2)

Therefore, from the above eq.(1) and (2), we get $x_1 = \pm 2\sqrt{6} = \dfrac{-k}{3}$, and thus, $k = \pm 6\sqrt{6}$.

(eg. 2) If the roots of the equation $ax^2 + bx + c = 0$ are α and β, the value of $\dfrac{(\alpha + \beta)^2}{\alpha\beta}$ in terms of a, b and c is?

Recall from the properties of the quadratic function, we have sum of $\alpha + \beta = \dfrac{-b}{a}$, $\alpha\beta = \dfrac{c}{a}$.

$\therefore \dfrac{(\alpha + \beta)^2}{\alpha\beta} = \dfrac{\left(\dfrac{-b}{a}\right)^2}{\dfrac{c}{a}} = \left(\dfrac{-b}{a}\right)^2 \cdot \dfrac{a}{c} = \dfrac{b^2}{a^2} \cdot \dfrac{a}{c} = \dfrac{b^2}{ac}$

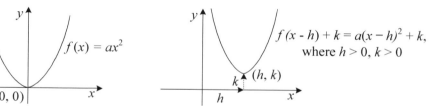

SAT II Math – Formula and Notes

Discriminant Testing:

For $D = b^2 - 4ac$,
- $D > 0$, then there are two real roots;
- $D = 0$, then one double roots;
- $D < 0$, there is no real roots.

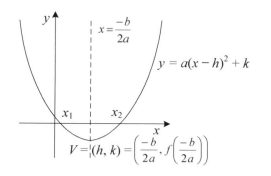
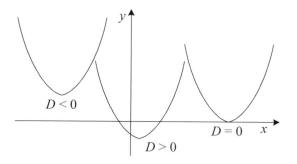

Note: Cubic function;

> Let $f(x) = ax^3 + bx^2 + cx + d$, then sum of the three roots, $x_1 + x_2 + x_3 = -\dfrac{b}{a}$
> and the product of the roots, $x_1 \cdot x_2 \cdot x_3 = -\dfrac{d}{a}$.

(eg. 1) If the roots of the equation $x^3 + px^2 + qx + r = 0$ are s_1, s_2 and s_3, then $s_1^2 + s_2^2 + s_3^2 = ?$

Recall from the cubic equation, $ax^3 + bx^2 + cx + d = 0$, the sum of 3 roots $x_1 + x_2 + x_3 = -\dfrac{b}{a}$,
$x_1x_2 + x_2x_3 + x_3x_1 = \dfrac{c}{a}$, $x_1 \cdot x_2 \cdot x_3 = -\dfrac{d}{a}$. In this problem, $a = 1$, $b = p$, $c = q$, $d = r$.
$\therefore s_1 + s_2 + s_3 = -p$, $s_1s_2 + s_2s_3 + s_3s_1 = q$
$\therefore s_1^2 + s_2^2 + s_3^2 = (s_1 + s_2 + s_3)^2 - 2(s_1s_2 + s_2s_3 + s_3s_1) = (-p)^2 - 2q = p^2 - 2q$.

2. Shifting and Translating of The Functions;

Let $y = f(x - h) + k$, represents the standard form of $f(x)$, then $f(x)$ has a horizontal translation of h unit, and a vertical shift of k unit. When $h > 0$, then move to the right, and $h < 0$, move to the left etc…

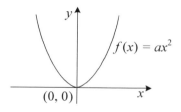
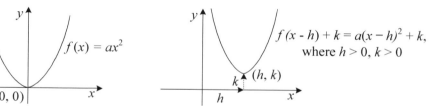

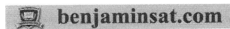

SAT II Math – Formula and Notes

(eg. 1) Suppose the graph of $f(x) = x^2 + x + 1$ is translated 2 units up and 1 units right.
If the resulting graph represents the graph of $g(x)$, what is the value of $g(-1.5)$?

$y = f(x - h) + k$, represents the standard form in which $f(x)$ has a horizontal translating of h unit,
and a vertical shifting of k unit. When $h > 0$, then move to the right, and $h < 0$, move to the left etc…
Now, $g(x)$ represents the graph of $f(x)$ with $h = 1$, and $k = 2$.
Therefore, $g(x) = f(x - 1) + 2$, or plugging in $(x - 1)$ for x, and adding 2,
we get $g(x) = [(x - 1)^2 + (x - 1) + 1] + 2 = (x - 1)^2 + (x - 1) + 3$.
Thus, $g(-1.5) = [(-2.5)^2 + (-2.5) + 3] = 6.75$.

(eg. 2) Suppose the graph of $f(x) = x^2 - 4x + 7$ is translated 2 units left and 3 unit down.
If the resulting graph represents $g(x)$, what is the value of $g(-0.5)$?

Since x is translated 2 units left, we get, $x \to x - (-2) = x + 2$. Also, y is 3 down, $y \to y - (-3) = y + 3$
$\therefore g(x) + 3 = f(x + 2)$, or $g(x) = f(x + 2) - 3 = (x + 2)^2 - 4(x + 2) + 7 - 3$
Now, $g(-\frac{1}{2}) = (\frac{3}{2})^2 - 4 \cdot \frac{3}{2} + 4 = \frac{9}{4} - 6 + 4 = \frac{1}{4}$

3. Polynomial and Rational Functions

A) The definition of even and odd function,

Even Function, if $f(-x) = f(x)$ or symmetric to y-axis, and
Odd Function, $f(-x) = -f(x)$ or symmetric to origin $(0, 0)$.

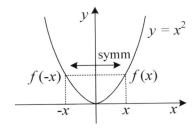

Even Function: $f(-x) = f(x)$
Even Function: symmetric to y-axis

Odd Function: $f(-x) = -f(x)$
Odd Function: symmetric to origin

(eg. 1) Which of the following is *not* an even function?

(A) $y = |x| + 1$ (B) $y = \cos x$ (C) $y = \log x^2$ (D) $y = x^2 + \tan x$ (E) $y = 4x^4 + 2x^2 + 3$

The definition of even function is, $f(-x) = f(x)$, or symmetric to y-axis.
In this problem, instead of using the definition, it would be better to use a graphing calculator
to see if any of the function given is <u>not</u> symmetric to y-axis.
The choice (D) $y = x^2 + \tan x$ is not symmetric to y-axis.

SAT II Math – Formula and Notes

(eg. 2)   If $f(x) = \sec x$ and $g(x) = 3x - 1$, which of the following is an even function (are even functions)?

I. $f(x) \cdot g(x)$
II. $f(g(x))$
III. $g(f(x))$

Since even function is symmetric to y-axis, using graphing utility,

let $f(x)g(x)$ as $y_1 = (\sec x)(3x - 1) = \dfrac{1}{\cos x} \cdot (3x - 1)$, and $f(g(x))$ as, $y_2 = \sec(3x - 1) = \dfrac{1}{\cos(3x-1)}$.

Also, $g(f(x))$ as, $y_3 = 3(\sec x) - 1 = \dfrac{3}{\cos x} - 1$.

Then, we find that (III) $g(f(x)) = 3(\sec x) - 1$ is the only one that is symmetric to y-axis, thus it is even function.

(eg. 3)   If $f(x) = 2x$ and $g(x) = x^2 + 1$, which of the following must be true?

    I. $f(x)g(x)$ is an odd function.
    II. $f(g(x))$ is an even function.
    III. $g(f(x))$ is an even function.

To be even function, it is symmetric to y-axis, and also to be odd function, it should be symmetric to origin $(0, 0)$. Now, by trying graphing utility for (I), (II) and (III),
- $y_1 = f(x)g(x) = 2x(x^2 + 1)$ shows symmetric to $(0, 0)$ ∴ True, odd function.
- (II) $y_2 = f(g(x)) = 2(x^2 + 1)$ shows symmetric to y-axis.   ∴ True, even function.
- (III) $y_3 = g(5(x)) = (2x)^2 + 1$ shows symmetric to y-axis.   ∴ True, even function.

B) Remainder Theorem and Factor Theorem:

- The remainder Theorem says, when $P(x)$ is divided by $(x - a)$, then the remainder becomes $P(a)$. For example, given polynomial function, $P(x) = 4x^4 + 3x^3 + 2x^2 - x - 1$, when this function is being divided by $(x + 1)$, we get the remainder, $P(x = (-)1) = P(-1) = 4(-1)^4 + 3(-1)^3 + 2(-1)^2 - (-1) - 1 = 3$.

- The Factor Theorem says, when $P(x)$ is divisible by $(x - a)$, then $P(x)$ must have a factor of $(x - a)$, and thus $P(a) = 0$.

(eg. 1)   Find the value of P if $(x + 1)$ is a factor of $f(x) = 3x^4 + 2x^3 - Px^2 - 2x + 1$.

    If $(x + 1)$ is a factor of $f(x)$, then $f(-1) = 0$.
    ∴ $f(-1) = 3 - 2 - p + 2 + 1 = 0$.
    ∴ $p = 4$

(eg. 2)   If n is an integer, what is the remainder when $-2x^{2n+1} - 3x^{2n} + 3x^{2n-1} + 4$ is divided by $x + 1$?

    Referring to our note on "remainder Theorem", where $p(a)$ = remainder, when $p(x)$ is divided by $(x - a)$.
    ∴ $p(-1) = -2(-1)^{2n+1} - 3(-1)^{2n} + 3(-1)^{2n-1} + 4 = 2 - 3 - 3 + 4 = 0$, which really means $(x + 1)$ is a factor of $p(x)$.

(eg. 3)   If 4 and -3 are both zeros of the polynomial $p(x)$, then a factor of $p(x)$ is?

    4 and $(-)3$ are both zeros of the polynomial $p(x)$.
    That means, by <u>Factor Theorem</u>, we get $p(4) = 0$ and $p(-3) = 0$.
    ∴ $p(x)$ must have $(x - 4)$ and $(x + 3)$ factors.
    ∴ $p(x) = (x - 4)(x + 3) = x^2 - x - 12$

SAT II Math – Formula and Notes

C) Horizontal and Vertical Asymptotes of the Rational Functions, $R(x) = F(x)/G(x)$;

- Vertical Asymptotes is the vertical line that the curve is approaching,
  but never cross over as $x \to a$ (There are some exceptional cases).
  In most cases, the rational function has its VA, when we find its denominator equals to zero.

- Horizontal Asymptotes is the horizontal line that the curve is approaching as $x \to \infty$.
  To find HA, we need to know 3 different cases. That is,

  1) When the degree of numerator = the degree of denominator,
     then HA is just the ratio of the coef. of the highest degree terms.
  2) When the degree of the Num < the degree of Denominator, then it is ZERO.
  3) When the degree of Num > the degree of Denom, then it is UNDEFINED.

- For example, let $y = \dfrac{x^2}{x^2 - 1}$, where $x \neq$ -1, 1.
  Then, since the degrees of numerator and denominator are both two, the HA is the limit value,
  which is the ratio of the leading coefficients. That is, our horizontal asymptote is $y = \dfrac{1}{1} = 1$.

  Now, to get the vertical asymptotes, where the function is undefined, we set its denominator equal zero.
  That is, $x^2 - 1 = 0$. Our vertical asymptotes are $x = 1$, $x = -1$.
  (Note: Be careful about "removable discontinuity"!!)
  In general, during Math IIC test when time management is very important, we may just use a graphing
  calculator, which is a much more quicker way to solve quickly during the exam.

  You may find from the graphing of $y = \dfrac{x^2}{x^2 - 1}$, two vertical asymptotes at $x = 1$, $x = -1$,
  and one horizontal asymptote at $y = 1$.

(eg 1) $y = \dfrac{x^2}{x^2 - 1}$

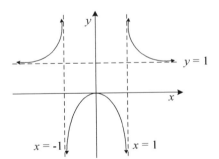

H.A. is, $\lim\limits_{x \to \infty} \dfrac{x^2}{x^2 - 1} = 1$

V.A. is, Denominator, $(x^2 - 1) = (x + 1)(x - 1) = 0$

$\therefore x = 1$ and $x = -1$

(eg 2) $y = \dfrac{x^2 - 3x + 2}{x^2 - 1} = \dfrac{(x-2)(x-1)}{(x+1)(x-1)} = \dfrac{x-2}{x+1} = \dfrac{x+1-3}{x+1} = 1 - \dfrac{3}{x+1}$

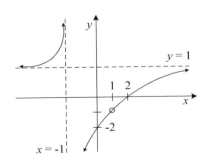

H.A. is, $y = 1$
V.A. is, $x = -1$
Removable Discontinuity is at $x = 1$

SAT II Math – Formula and Notes

(eg 3) $y = \dfrac{x^2+2x+2}{x+1} = (x + 1) + \dfrac{1}{x+1}$

H.A. None
V.A. $x = -1$
Oblique asymptotes: $y = x + 1$

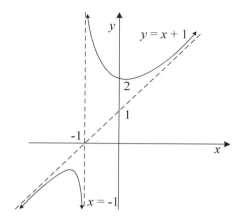

4. Exponential Functions

Problems involve such as calculating interest rate, population growth or any exponential growth or decay, we have the following formula in two different cases.

- Discrete case: $A = A_0(1 + \dfrac{r}{n})^{nt}$, where r is annual rate, A_0 is an initial amount,

- n is number of times that annual rate r is compounded, and t is time period.

- Continuous case: $A = A_0 e^{rt}$, or $C = C_0 e^{kt}$, where positive k is for growth, and negative k for decay.

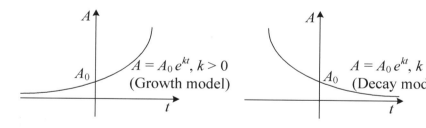

(eg. 1) If an initial investment of $2,000 in a savings account is compounded continuously at an annual rate of 6 percent, how many years will it take the investment to be worth approximately $5,000?

Notice that the saving is compounded <u>continuously</u> at $r = 6\%$ or, $r = 0.06$ annually.
In this problem, we choose the continuous case, because it is compounded continuously. $y = A_0 e^{rt}$
where $A_0 = 2000$, $r = 0.06$, $y = 5,000$. This leads us to our eq. $5000 = 2000 \times e^{0.06t}$, or $\dfrac{5000}{2000} = e^{0.06t}$.

Applying natural log, ln on both sides, we get $\ln \dfrac{5}{2} = 0.06t$. Therefore, $t = \dfrac{\ln \frac{5}{2}}{0.06} = 15.3$

SAT II Math – Formula and Notes

(eg. 2)   When a certain radioactive element decays, the amount that exists at any time t can be calculated by the function $A(t) = A_0 e^{\frac{-t}{1000}}$, where A_0 is the initial amount and t is the elapsed time in years. How many years would it take for an initial amount of 800 milligrams of this element to decay to 200 milligrams?

$$200 = 800 \cdot e^{\frac{-t}{1000}}$$

$$\therefore \frac{200}{800} = \frac{1}{4} = e^{\frac{-t}{1000}}$$

Now, applying "ln" on both sides, we get $\ln \frac{1}{4} = \ln e^{\frac{-t}{1000}} = \frac{-t}{1000}(\ln e) = \frac{-t}{1000}$ (1).

$$\therefore t = -1000 \cdot \ln \frac{1}{4} = 1{,}386$$

5. The inverse of a function is defined by $f^{-1}(x)$, for which,

- $f(x)$ must be one to one function.

- $f(x)$ and $f^{-1}(x)$ are symmetric to $y = x$ line

- To get $f^{-1}(x)$, we need to switch x into y, and y into x.

- The Domain and Range of $f(x)$ become the Range and the Domain of $f^{-1}(x)$ by interchanging respectively.

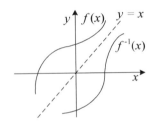

(eg. 1)   What is the inverse of the function of $f(x) = \frac{1}{\sqrt{x-1}}$, $x > 1, y > 0$?

Here, $f(x) = \frac{1}{\sqrt{x-1}} = y$. Now, by interchanging x and y, we get $\frac{1}{\sqrt{y-1}} = x$ or $\frac{1}{\sqrt{(y-1)^2}} = x^2$.

That is $\frac{1}{y-1} = x^2$ or $y - 1 = \frac{1}{x^2}$, which results in $y = \frac{1}{x^2} + 1 = \frac{1+x^2}{x^2}$ with $x > 0, y > 1$.

6. Properties of log function:

1) $\log x \cdot y = \log x + \log y$

2) $\log \frac{x}{y} = \log x - \log y$

3) $\log x^m = m \cdot \log x$

4) $\log_x y = \frac{\log y}{\log x} = \frac{\ln y}{\ln x}$

5) $\log_x x = 1$, $\log_x 1 = 0$

6) $y = \log_a x$, if and only if $x = a^y$

SAT II Math – Formula and Notes

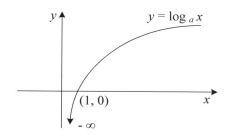

(eg. 1) If $\log_a 9 = 2$, then what is a?

$\log_a 9 = 2 \rightarrow 9 = a^2$ by the definition of logarithm.
$\therefore a = 3$.

(eg. 2) Approximate $\log_8 13$?

$\log_A B = \dfrac{\log B}{\log A}$

$\therefore \log_8 13 = \dfrac{\log 13}{\log 8} = 1.23$

(eg. 3) If $\log_9 p = 3$ and $\log_3 q = 4$, p expressed in terms of q is ?

$\log_9 p = 3 \rightarrow p = 9^3 = (3^2)^3 = 3^6$, $\log_3 q = 4 \rightarrow q = 3^4$

$\therefore p = 3^6 = (3^4)^{\frac{3}{2}} = (q)^{\frac{3}{2}}$

7. The formula for the distance D between a point (x, y) and a line $Ax + By + C = 0$;

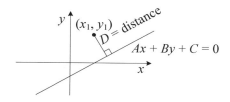

$D = \dfrac{|Ax_1 + By_1 + C|}{\sqrt{A^2 + B^2}}$

(eg. 1) If point $P(k, 2k)$ is 3 units from line $3x + 4y = 7$, then k equals ?

The formula for the distance D between a point (x, y) and a line $Ax + By + C = 0$ is, $D = \dfrac{|Ax_1 + By_1 + C|}{\sqrt{A^2 + B^2}}$.

Using this formula with $(x_1, y_1) = (k, 2k)$ and $A = 3$, $B = 4$, $C = (-)7$,

we get $D = \dfrac{|3(k) + 4(2k) - 7|}{\sqrt{3^2 + 4^2}} = 3$, or $(11k - 7) = \pm 15$.

Thus, $11k = 7 \pm 15$, or $k = 2$, $k = (-)0.73$.

SAT II Math – Formula and Notes

8. Arithmetic Series and Geometric Series:

- For Arithmetic series,

$$a_n = a_1 + (n-1)d, \quad d = a_n - a_{n-1}, \quad s_n = \frac{a_1 + a_n}{2} \cdot n, \text{ where } d \text{ is the common difference.}$$

(eg. 1) If the 10th term of an arithmetic sequence is 15 and the 20th term of the sequence is 100, what is the first term of the sequence?

Referring to the previous note on "Sequence and Series", where $a_n = a_1 + (n-1)d$,
$a_{10} = a_1 + 9d = 15$, $a_{20} = a_1 + 19d = 100$
$\therefore a_{20} - a_{10} = 10d = 85$ $\therefore d = 8.5$
$\therefore a_1 = 15 - 9d = 15 - 9(8.5) = (-)61.5$

- For Geometric Series;

$$a_n = a_1 r^{n-1}, \quad r = \frac{a_n}{a_{n-1}}, \quad s_n = a_1 \cdot \frac{1 - r^n}{1 - r}$$

(eg. 1) If the 3rd term of a geometric progression is $\sqrt[3]{a}$ and the 6th term is $\sqrt[3]{a^2}$, what is the 12th term of the progression?

Recall the Geometric series with the following formulas:

$$a_n = a_1 r^{n-1}, \quad r = \frac{a_n}{a_{n-1}}, \quad s_n = a_1 \frac{1 - r^n}{1 - r}$$

And also, the infinite series, $s = a_1 \frac{1}{1-r}$, $|r| < 1$. Here, $a_3 = a^{\frac{1}{3}}$ and $a_6 = a^{\frac{2}{3}}$.

$\therefore \frac{a_6}{a_3} = \frac{a_1 r^5}{a_1 r^2} = r^3$, which is, $\frac{a^{\frac{2}{3}}}{a^{\frac{1}{3}}} = r^3$, or $a^{\frac{1}{3}} = r^3 \rightarrow r = a^{\frac{1}{9}}$. Therefore, $a_{12} = a_6 \cdot r^6 = a^{\frac{2}{3}} \cdot \left(a^{\frac{1}{9}}\right)^6 = a^{\frac{2}{3}} \cdot a^{\frac{2}{3}} = a^{\frac{4}{3}}$.

- For Infinite Series,

$$s = a_1 \left(\frac{1}{1-r}\right), \ |r| < 1$$

(eg. 1) Find the product of an infinite number of terms: $5^{\frac{1}{3}} \times 5^{\frac{2}{9}} \times 5^{\frac{4}{27}} \times 5^{\frac{8}{81}} \times \ldots$

$5^{\frac{1}{3}} \times 5^{\frac{2}{9}} \times 5^{\frac{4}{27}} \times 5^{\frac{8}{81}} \ldots = 5^{\frac{1}{3}(1 + \frac{2}{3} + \frac{4}{9} + \frac{8}{27} + \cdots)} = 5^{\frac{1}{3}(1 + \frac{2}{3} + \left(\frac{2}{3}\right)^2 + \left(\frac{2}{3}\right)^3 + \cdots)}$

Now, consider a geometric series: $1 + \frac{2}{3} + \left(\frac{2}{3}\right)^2 + \left(\frac{2}{3}\right)^3 + \ldots$ which is an infinite series with $a_1 = 1$, $r = \frac{2}{3}$

$\therefore$ The sum of the series, $S = \frac{1}{1 - \frac{2}{3}} = 3$

$\therefore 5^{\frac{1}{3}(1 + \frac{2}{3} + \left(\frac{2}{3}\right)^2 + \left(\frac{2}{3}\right)^3 + \cdots)} = 5^{\frac{1}{3}(3)} = 5^1 = 5$

SAT II Math – Formula and Notes

(eg. 2) What is the sum of the infinite geometric series $1 + \frac{1}{2} + \frac{1}{4} + \frac{1}{8} + \frac{1}{16} + \frac{1}{32} + \dots$?

For a given Geometric Sequence; $a_1, a_2, a_3, \dots, a_n, \dots$, we get $a_n = a_1 r^{n-1}$ and $s_n = a_1 \frac{1-r^n}{1-r}$.

Furthermore, for an infinite series, we have the sum, $s = a_1 \frac{1}{1-r}, |r| < 1$.

Here, since our $r = \frac{1}{2}$, and $|r| < 1$, we get $s = (1) \frac{1}{1-\frac{1}{2}} = 2$

9. Logic:

There are 4 different logic cases;
Let's make a statement. If p, then q. Using mathematical symbol for this statement is, $p \rightarrow q$.

- Negation: If p, then <u>NOT</u> q, or, $p \rightarrow \sim q$, where "$\sim$" sign represents "NOT".
- Inverse: If <u>NOT</u> p, then <u>NOT</u> q, or $\sim p \rightarrow \sim q$.
- Converse: If q, then p, or $q \rightarrow p$.
- Contra Positive: If <u>NOT</u> q, then <u>NOT</u> p, or $\sim q \rightarrow \sim p$.

Of all these, the Contra Positive is the most important one to remember. Here, the Contra Positive is exactly the same statement as the original statement. Often, if we cannot prove the original statement, then we may use the logic of Contra Positive to prove the original statement.

(eg. 1) The inverse of $p \rightarrow \sim q$ is equivalent to ?

The inverse of $p \rightarrow \sim q$ is equivalent to negation on both, or $\sim p \rightarrow \sim(\sim q)$, which is $\sim p \rightarrow q$, because $\sim(\sim q) = q$ itself. However, we don't have any answer choice for this. Therefore, making Contra Positive of $\sim p \rightarrow q$ by switching $\sim p$ and q, and negating both, we get $\sim q \rightarrow \sim(\sim p)$ or $\sim q \rightarrow p$.

(eg. 2) Which of the following statements is logically equivalent to: "If she exercises, she will pass the test."

(A) She passed the test; therefore, she exercised.
(B) She did not exercise; therefore, she will not pass the test.
(C) She did not pass the test; therefore she did not exercise.
(D) She will pass the test only if she exercises.
(E) None of the above.

In this problem, "logically equivalent" means to find "contra positive". Referring to our previous note on "logic", the contra positive of the statement $p \rightarrow q$ is $\sim q \rightarrow \sim p$, where "$\sim$" means negation.
∴ "If she exercise, she will pass the test." becomes "If she does <u>not</u> pass the test, then she did <u>not</u> exercise."

(eg. 3) An indirect proof of the statement "If $x < 0$, then $\sqrt{x}$ is <u>not</u> a real number" could begin with the assumption that

(A) $x = 0$
(B) $x > 0$
(C) $\sqrt{x}$ is real number
(D) $\sqrt{x}$ is not a real number
(E) x is nonnegative

"An <u>indirect proof</u>" really means "Contra Positive" in logic, which means, the statement "if <u>P, then Q</u>", implies the contra positive statement of "if <u>not</u> Q, then <u>not</u> P". Applying this, we get, if $\sqrt{x}$ is a real number, then $x \geq 0$.

SAT II Math – Formula and Notes

10. Binomial Expansion,

$(A + B)^n = \sum_{r=0}^{n} {}_nC_r \, A^{(n-r)}B^r$

(eg. 1) When $(a + b)^4$ is expanded, what is the coefficient of the third term?

Refer to Binomial Expansion,

$$(a + b)^n = \sum_{r=0}^{n} {}_nC_r \cdot a^{n-r} b^r$$

Using $n = 4$, $r = 2$, we get the coefficient of the third term, ${}_4C_2 = 6$.

(eg. 2) Which of the following could be a term in the expansion of $(a - b)^8$?

(A) $56a^5b^3$

(B) $-56a^5b^3$

(C) $56a^6b^2$

(D) $-56a^6b^2$

(E) $-56a^2b^6$

Referring to the previous note on "Binomial Expansion".
$(a + b)^n = \sum_{r=0}^{n} {}_nC_r \cdot a^{n-r} b^r$, $(a - b)^8 = \sum_{r=0}^{8} {}_8C_r \cdot a^{8-r}(-b)^r$.
Here, let $r = 3$, then we get ${}_8C_3 a^5(-b)^3 = -56a^5b^3$.

11. Trigonometry

A) 6 Basic Trig Functions;

$\sin \theta = \dfrac{\text{opp}}{\text{hyp}} = \dfrac{a}{c}$, $\cos \theta = \dfrac{\text{adj}}{\text{hyp}} = \dfrac{b}{c}$, $\tan \theta = \dfrac{\text{opp}}{\text{adj}} = \dfrac{a}{b}$,

$\csc \theta = \dfrac{1}{\sin \theta}$, $\sec \theta = \dfrac{1}{\cos \theta}$, $\cot \theta = \dfrac{1}{\tan \theta}$

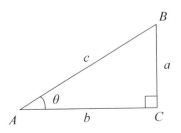

SAT II Math – Formula and Notes

1. $y = \sin\theta$,

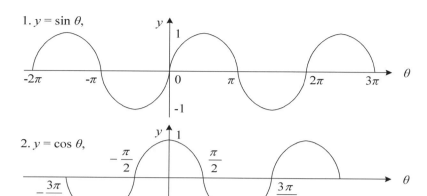

2. $y = \cos\theta$,

3. $y = \tan\theta$,

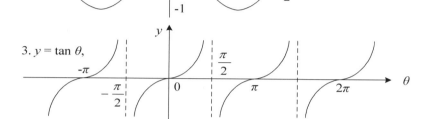

B) The property of the angle sum and difference

- $\sin(A \pm B) = \sin A \cos B \pm \cos A \sin B$
- $\cos(A \pm B) = \cos A \cos B \mp \sin A \sin B$
- $\tan(A \pm B) = \dfrac{\tan A \pm \tan B}{1 \mp \tan A \tan B}$
- $\sin 2A = 2\sin A \cos A$
- $\cos 2A = \cos^2 A - \sin^2 A = 1 - 2\sin^2 A = 2\cos^2 A - 1$
- $\sin^2 A + \cos^2 A = 1$
- $1 + \tan^2\theta = \sec^2\theta$

(eg. 1) $\cos(150° + x) - \cos(150° - x)$ equals

(A) $\sqrt{2}\sin x$
(B) $-\sin x$
(C) $\sqrt{3}\cos x$
(D) $\sqrt{2}\cos x$
(E) $\sqrt{3}\sin x$

Refer to $\cos(A + B) = \cos A \cdot \cos B - \sin A \cdot \sin B$.
Then $\cos(150° + x) - \cos(150° - x)$
$= (\cos 150° \cdot \cos x - \sin 150° \cdot \sin x) - (\cos 150° \cdot \cos x + \sin 150° \cdot \sin x)$
$= (-)2 \cdot \sin 150° \cdot \sin x = (-)2 \cdot \dfrac{1}{2} \sin x = -\sin x$
But, for this problem, I strongly recommend to use a calculator by letting $x = 30°$ or some other angle.
Then $\cos(150° + 30°) - \cos(150° - 30°) = \cos 180° - \cos 120° = -1 - (-\dfrac{1}{2}) = -\dfrac{1}{2}$
Try answer choice (A) $\sqrt{2}\sin 30° = \sqrt{2}\dfrac{1}{2} = \dfrac{1}{\sqrt{2}}$ $\therefore$ No!,
(B) $-\sin 30° = -\dfrac{1}{2}$ $\therefore$ Yes

SAT II Math – Formula and Notes

(eg. 2) If $\sin x = \dfrac{3}{5}$ and $\dfrac{\pi}{2} \leq x \leq \pi$, then $\tan 2x = ?$

$\sin x = \dfrac{3}{5}$

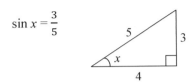

Now, from the triangle shown above, we get $\tan x = \dfrac{3}{4}$.

But our domain $\dfrac{\pi}{2} \leq x \leq \pi$ implies it is in the 2nd quadrant, where $\tan x = -\dfrac{3}{4}$.

Now, $\tan 2x = \dfrac{2\tan x}{1-\tan^2 x} = \dfrac{2\left(-\dfrac{3}{4}\right)}{1-\left(-\dfrac{3}{4}\right)^2} = \dfrac{-\dfrac{6}{4}}{\dfrac{7}{16}} = -\dfrac{6}{4} \times \dfrac{16}{7} = -\dfrac{24}{7}$

C) Heron's formula to find the area, given 3 sides;

$$\text{Area} = \sqrt{s(s-a)(s-b)(s-c)}, \text{ where } s = \frac{a+b+c}{2}$$

(eg. 1) The area of a triangle with sides 4, 5, and 8 is

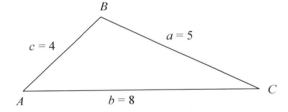

Here, we are given the lengths of 3 sides only, without any angle information.

Therefore, we use Heron's formula to find the area, that is $s = \dfrac{a+b+c}{2} = \dfrac{4+5+8}{2} = 8.5$

$\text{Area} = \sqrt{s(s-a)(s-b)(s-c)} = \sqrt{8.5(8.5-4)(8.5-5)(8.5-8)} = 8.18$

D) Law of Sine and Law of Cosine

- The Law of Sine: $\dfrac{\sin A}{a} = \dfrac{\sin B}{b} = \dfrac{\sin C}{c}$

- The law of Cosine: $c^2 = a^2 + b^2 - 2ab \cdot \cos C$

- Area with 2 side, and angle θ: $\text{Area} = \dfrac{1}{2} \cdot bc \cdot \sin \theta$

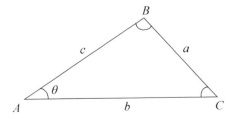

SAT II Math – Formula and Notes

(eg. 1)   What is the measure of the smallest angle in a right triangle with sides of lengths 5, 12 and 13?

Referring to the topics of Trig. Functions previously, we may use the Law of Cosine,
$c^2 = a^2 + b^2 - 2ab \cdot \cos C$. Here, the shortest side is 5.
$\therefore 5^2 = 12^2 + 13^2 - 2 \cdot 12 \cdot 13 \cdot \cos C$.
$\therefore C = \cos^{-1}\left(\dfrac{12^2 + 13^2 - 5^2}{2 \cdot 12 \cdot 13}\right) = 22.62°$

(eg. 2)   What is the degree measure of the angle x of a triangle that has sides of length of 5 and 7 as shown in the figure (1)?

Using the Law of sine, we get $\dfrac{5}{\sin 30°} = \dfrac{7}{\sin x°}$

$\therefore \sin x = 7 \cdot \dfrac{\sin 30°}{5} = \dfrac{7}{10}$ $\therefore x = \sin^{-1}\dfrac{7}{10} = 44.43°$

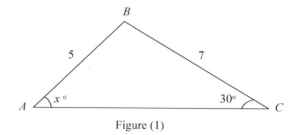

Figure (1)

(eg. 3)   In $\triangle ABC$, $a = 2$, $b = 5$, $\angle C = 60°$. What is the area of $\triangle ABC$?

Area of $\triangle ABC = \dfrac{1}{2} \cdot a \cdot b \cdot \sin c°$

$= \dfrac{1}{2}(2)(5) \cdot \sin 60° = \dfrac{5\sqrt{3}}{2}$.

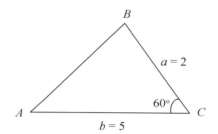

(eg. 4)   In figure 37, what is the approximate area of triangle ABC?

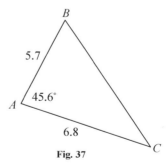

Fig. 37

The area of $\triangle ABC$ is $\dfrac{1}{2}bh = \dfrac{1}{2}bc \sin \theta$, where $h = c \cdot \sin \theta$.
Since we have $b = 6.8$, $c = 5.7$ and $\theta = 45.6°$,
we have area $\triangle ABC = \dfrac{1}{2} \cdot (6.8)(5.7) \sin 45.6° = 13.85$.

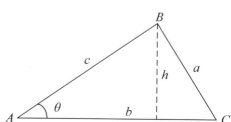

SAT II Math – Formula and Notes

E) <u>For a general form</u>: $y = a \cos b(x - h) + k$,

we have, $| a | =$ Amplitude, $b = \#$ of frequency, $\dfrac{2\pi}{b} =$ period,
$h =$ units of translation horizontally, $k =$ units of shifting vertically.

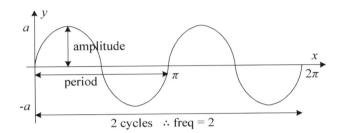

(eg. 1)   What is the period of the graph of the function $y = \cos^2 x - \sin^2 x$?

Note that $\cos 2x = \cos^2 x - \sin^2 x$.
So, our equation becomes, $y = \cos^2 x - \sin^2 x = \cos 2x$, and thus, the period $p = \dfrac{2\pi}{2} = \pi$.

(eg. 2)   What is the period of the curve whose equation is $y = 4\sin x \cdot \cos x$?

$y = 4\sin x \cdot \cos x = 2(2\sin x \cdot \cos x)$.
 But $2\sin x \cdot \cos x = \sin 2x$.
$\therefore y = 2 \cdot \sin 2x$.
$\therefore p = 2\,\dfrac{\pi}{2} = \pi = 180°$

F) <u>Cofunctions</u>

$\sin \theta = \cos (90° - \theta) = \cos (\dfrac{\pi}{2} - \theta)$,   (ie) $\sin 30° = \cos 60°$

$\tan \theta = \cot (90° - \theta) = \cot (\dfrac{\pi}{2} - \theta)$,   (ie) $\tan 15° = \cot 75°$

$\sec \theta = \csc (90° - \theta) = \csc (\dfrac{\pi}{2} - \theta)$,   (ie) $\sec 45° = \csc 45°$

(eg. 1)   If A is the degree measure of an acute angle and $\sin A = 0.25$, then $\sin A \cdot \cos (90° - A)$?

Recall the property of <u>Cofunction</u>. $\sin (90° - A) = \cos A$
(eg) $\sin 30° = \cos 60°$, $\tan 25° = \cot 65°$, $\sec 15° = \csc 75°$, ...
Therefore, $\sin A \cdot \cos (90° - A) = (\sin A)(\sin A) = (0.25)(0.25) = 0.0625$

(eg. 2)   If $\tan A = \cot B$, which of the following must be true?

(A) $A = B$
(B) $A + B = 0$
(C) $A + B = 90°$
(D) $A + B = 180°$
(E) $A - B = 180°$

Since $\tan A = \cot B$, by the property (2), we must get $A + B = 90°$.

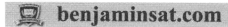

SAT II Math – Formula and Notes

G) Sign changes by Quadrants; all seniors take calculus… by I-II-III-IV Qtr

$$
\begin{array}{c|c}
\text{sin (+)} & \text{all (+)} \\
\hline
\text{tan (+)} & \text{cos (+)}
\end{array}
$$

(eg. 1) If $(\csc x)(\cot x) < 0$, which of the following must be true?

    I. $\tan x < 0$
    II. $(\sin x)(\cos x) < 0$
    III. x is in the second or third quadrant

$$
\begin{array}{c|c}
\text{(II)} & \text{(I)} \\
\text{sin (+)} & \text{All} \\
\text{cos (-)} & \\
\text{tan (-)} & \\
\hline
\text{(III)} & \text{(IV)} \\
\text{tan (+)} & \text{cos (+)} \\
\text{cos (-)} & \text{tan (-)} \\
\text{sin (-)} & \text{sin (-)}
\end{array}
$$

Remember that $\csc x = \dfrac{1}{\sin x}$ and $\cot x = \dfrac{1}{\tan x}$. Therefore, $(\csc x)(\cot x) = \dfrac{1}{\sin x} \times \dfrac{1}{\tan x} < 0$, which implies that the signs of $\sin x$ and $\tan x$ must be the opposite signs each other. Referring to the sign chart by quadrant, we have $\sin x$ and $\tan x$ have opposite signs each other in (II) and (III) quadrants.

(eg. 2)

$$
\begin{array}{c|c}
\text{II} & \text{I} \\
\hline
\text{III} & \text{IV}
\end{array}
$$

If $\sin \theta \cdot \cos \theta < 0$ and $\sin \theta \cdot \tan \theta > 0$, then θ must be in which quadrant in the figure above?

$\sin \theta \cdot \cos \theta < 0$ ∴ $\sin \theta$ and $\cos \theta$ must have opposite signs! ∴ II or IV
$\sin \theta \cdot \tan \theta > 0$ ∴ $\sin \theta$ and $\tan \theta$ must have the same signs. ∴ I or IV
Now, IV is the only quadrant that satisfy both!!

SAT II Math – Formula and Notes

12. De Moivre's Theorem;

Let the complex number, $Z = a + bi = r(\cos \theta + i \sin \theta)$, where $r = \sqrt{a^2 + b^2}$, $\theta = \tan^{-1} \dfrac{b}{a}$.

Then, $Z^n = (a + bi)^n = [r(\cos \theta + i \sin \theta)]^n = r^n[\cos (n\theta) + i \sin (n\theta)]$.

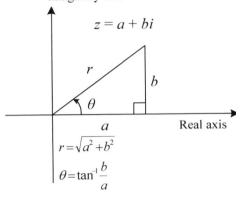

(eg. 1) The length of the vector that could correctly be used to represent in the complex plane is the magnitude of a vector,

It is given by $|z| = |a + bi| = \sqrt{a^2 + b^2}$.

Therefore, when z is given by $-\sqrt{7} + 2i$, $|z| = |-\sqrt{7} + 2i| = \sqrt{\left(-\sqrt{7}\right)^2 + 2^2} = \sqrt{11}$.

(eg. 2) What is the inverse of $f(x) = \dfrac{1}{\sqrt{-x^2+5}}$?

$y = f(x) = \dfrac{1}{\sqrt{-x^2+5}}$.

To get the inverse $f^{-1}(x)$ by switching x and y's, we get $x = \dfrac{1}{\sqrt{-y^2+5}}$ or $x^2 = \dfrac{1}{-y^2+5}$.

Now, by reciprocal of both sides, $\dfrac{1}{x^2} = -y^2 + 5 \rightarrow y^2 = 5 - \dfrac{1}{x^2}$, or $y = \sqrt{5 - \dfrac{1}{x^2}} = f^{-1}(x)$.

(eg. 3) Write $\left(\sqrt{2}(\cos 15° + i \sin 15°)\right)^3$ in the form $a + bi$.

Referring to De Moivres Theorem for the Complex Number, let the complex number,

$Z = a + bi = r(\cos \theta + i \sin \theta)$, where $r = \sqrt{a^2 + b^2}$, $\theta = \tan^{-1} \dfrac{b}{a}$.

Then, $Z^n = (a + bi)^n = [r(\cos \theta + i \sin \theta)]^n = r^n[\cos (n\theta) + i \sin (n\theta)]$.

Using this formula, we get $\left(\sqrt{2}(\cos 15° + i \sin 15°)\right)^3 = \left(\sqrt{2}\right)^3 \{\cos(3 \times 15°) + i \sin(3 \times 15°)\}$

$= 2\sqrt{2} (\cos 45° + i \sin 45°) = 2\sqrt{2} \left(\dfrac{1}{\sqrt{2}} + i \cdot \dfrac{1}{\sqrt{2}}\right) = 2 + 2i$

SAT II Math – Formula and Notes

13. Properties of vector-operation:

Let vector $\vec{v} = (v_1, v_2)$ and $\vec{u} = (u_1, u_2)$, then

1) $\vec{u} \pm \vec{v} = (u_1 \pm v_1, u_2 \pm v_2)$

2) $c\vec{u} = (cu_1, cu_2)$

3) magnitude or length of $\vec{u}$ or $|\vec{u}| = \sqrt{u_1{}^2 + u_2{}^2}$

4) Unit vector $\vec{w}$ of $\vec{v}$ is, $\vec{w} = \dfrac{\vec{v}}{|\vec{v}|} = \dfrac{(v_1, v_2)}{\sqrt{v_1{}^2 + v_2{}^2}}$

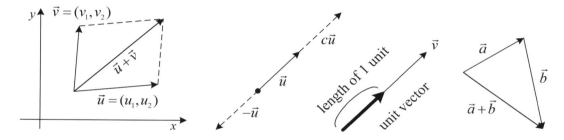

(eg. 1)   If vector $\vec{v} = (\sqrt{2}, \sqrt{3})$ and vector $\vec{u} = (1, -1)$, find the value of $|\sqrt{3}\vec{v} + 3\vec{u}|$.

Now, given $\vec{v} = (\sqrt{2}, \sqrt{3})$, $\vec{u} = (1, -1)$.

We get, $\sqrt{3}\vec{v} = (\sqrt{3}\sqrt{2}, \sqrt{3}\sqrt{3}) = (\sqrt{6}, 3)$ and $3\vec{u} = (3, -3)$, such that $\sqrt{3}\vec{v} + 3\vec{u} = (3 + \sqrt{6}, 0)$.

Therefore, using the property (3), we get $|\sqrt{3}\vec{v} + 3\vec{u}| = \sqrt{\left(3 + \sqrt{6}\right)^2 + 0^2} = 3 + \sqrt{6}$.

14. Probability and Combination;

- The total possible number of arrangement on n objects is $n!$

  (eg)   (A, B, C) = 3!, because A-B-C, A-C-B, B-A-C, B-C-A, C-A-B, C-B-A, total of 6 = 3!

  (A, B, C, D) = 4!

  (A, B, C, D, E) = 5! ... etc.

- The total number of arrangement on n objects, but with the repetition of k and r objects is $\dfrac{n!}{k!r!}$

  (eg)   COFFEE $\rightarrow \dfrac{6 \text{ letters}}{(2 \text{ F}'\text{s})(2 \text{ E}'\text{s})} = \dfrac{6!}{2!2!}$

  Minimum $\rightarrow \dfrac{7 \text{ letters}}{(3 \text{ m}'\text{s})(2 \text{ i}'\text{s})} = \dfrac{7!}{3!2!}$

- The total number of arrangement on n objects, but with k group with n_1 and n_2, … members in them is $k!(n_1!)(n_2!)\ldots$

  (eg)   A, B, C, D, E into 2 grouping of (A, B), (C, D, E) will be

  (2! grouping) $\times$ (2! on A, B) $\times$ (3! on C, D, E) = (2!)(2!)(3!)

SAT II Math – Formula and Notes

- Permutation is, with a consideration of "Ordering", or $_nP_r = \dfrac{n!}{(n-r)!}$

    (eg)   Suppose I choose 3 best scored students out of 10 students in my Math IIc class, and then buy them lunch,
    steak for the best, hamburger for the second best, and tacos for the third best.
    Then, this will be a permutation, because "Order" matters by having different lunches with their rank,
    thus $_{10}P_3 = \dfrac{10!}{(10-3)!} = \dfrac{10!}{7!}$

- Combination is, without a consideration of "Ordering", or $_nC_r = \dfrac{n!}{(n-r)!\,r!}$

    (eg)   Suppose that I choose 3 best scored students out of 10 students in my Math IIc class, and then buy them
    lunch, but this time, all hamburgers. Then, in this case, "order" does not matter, because they all eat
    hamburgers anyway, regardless of their rank. Thus, it is Combination, $_{10}C_3 = \dfrac{10!}{(10-3)!3!} = \dfrac{10!}{7!3!}$.
    Now, notice that $_{10}C_3$ and $_{10}P_3$ are exactly the same, except that $_{10}C_3$ is being divided by 3!,
    while $_{10}P_3$ is not. That is because in $_{10}C_3$ case, those 3 students selected will have the same lunch,
    hamburger – hamburger – hamburger, or H - H - H → repetition of 3, therefore, being divided by 3!.

15. Conic Sections:

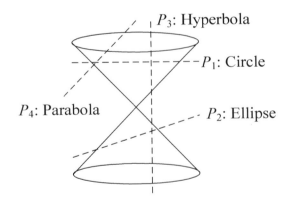

P_3: Hyperbola

P_1: Circle

P_4: Parabola

P_2: Ellipse

1) **Circle:** Let P_1 be the plane that cuts the cone in a horizontal direction.

Then, it creates a circle, formed by its intersection.

Given $(x-h)^2 + (y-k)^2 = r^2$,

- center $= (h, k)$
- radius $= r$

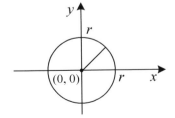

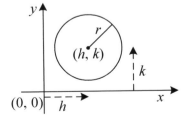

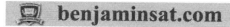
SAT II Math – Formula and Notes

(eg. 1) Which ordered number pair represents the center of the circle $x^2 + y^2 + 12x - 8y = 0$?

To get the center of this circle, we need to modify this eq. into a <u>complete square</u> form.
That is, $x^2 + 12x + y^2 - 8y = 0$ becomes $(x + 6)^2 + (y - 4)^2 = 52$.
Therefore, we get the center $(-6, 4)$.

(eg. 2) The circumference of circle $x^2 + y^2 - 6x - 4y - 36 = 0$ is

$x^2 - 6x + y^2 - 4y = 36 \rightarrow$ making a complete square form, $(x - 3)^2 + (y - 2)^2 = 36 + 9 + 4 = 49 = 7^2$
∴ center $c = (3, 2)$, radius $r = 7$
∴ circumference $c = 2\pi r = 2\pi(7) = 14\pi = 43.98$

2) Ellipse: Let P_2 be the plane that cuts the cone in <u>a tilted direction</u>.

Then, it creates an elliptical figure, formed by intersection.

Given $\dfrac{(x-h)^2}{a^2} + \dfrac{(y-k)^2}{b^2} = 1$

- center = (h, k)
- length of Major axis = $2a$, when $a > b$
- length of Minor axis = $2b$
- vertex = $(\pm a + h, o + k)$
- foci = $(\pm c + h, o + k)$, where $c^2 = a^2 - b^2$
- $d_1 + d_2 = K = 2a$

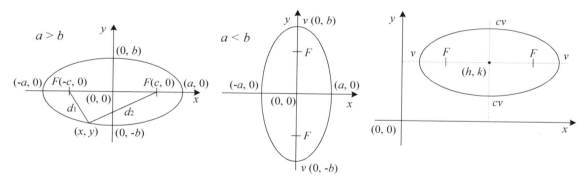

(eg. 1) What is the length of the major axis of the ellipse whose equation is $10x^2 + 6y^2 = 15$?

$10x^2 + 6y^2 = 15 \rightarrow \dfrac{10x^2}{15} + \dfrac{6y^2}{15} = 1$

$\rightarrow \dfrac{x^2}{\frac{15}{10}} + \dfrac{y^2}{\frac{15}{6}} = 1 \rightarrow \dfrac{x^2}{(\sqrt{1.5})^2} + \dfrac{y^2}{(\sqrt{2.5})^2} = 1$

∴ $a = \sqrt{1.5}$, $b = \sqrt{2.5}$

∴ The length of major axis = $2b = 2\sqrt{2.5} = 3.16$

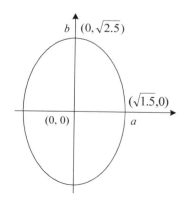

SAT II Math – Formula and Notes

(eg. 2) If $3(x + 1)^2 + 10(y - 2)^2 = 48$ is graphed the sum of the distances from any point on the curve to the two foci is ?

$$3(x + 1)^2 + 10(y - 2)^2 = 48 \rightarrow \frac{(x+1)^2}{16} + \frac{(y-2)^2}{4.8} = 1 \quad \therefore a = 4, b = \sqrt{4.8}$$

$\therefore$ The length of major axis = $2a = 8$, which is equal to the sum of the distances between two fixed points to a point on the curve.

3) Hyperbola: Let P_3 be the plane that cuts the cone in a vertical direction.

Then, it creates a hyperbolic figure with 2 curves, formed by its intersection.

Given $\frac{(x-h)^2}{a^2} - \frac{(y-k)^2}{b^2} = 1$, or $\frac{(y-k)^2}{b^2} - \frac{(x-h)^2}{a^2} = 1$

- center = (h, k)
- vertex = $(\pm a + h, o + k)$, or $(0 + h, \pm b + k)$
- asymptotes: $y - k = \pm \frac{b}{a}(x - h)$, where the slope $\frac{b}{a}$ is $\frac{rise}{run}$.
- foci = $(\pm c + h, o + k)$ or $(0 + h, \pm c + k)$, where $c^2 = a^2 + b^2$
- $d_1 - d_2 = K = 2a$

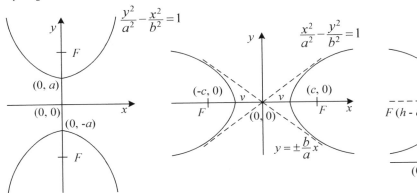
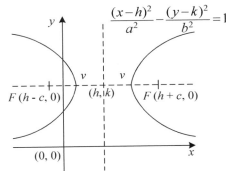

(eg. 1) When drawn on the same set of axes, the graphs of $x^2 - y^2 = 4$ and $4(x - 1)^2 + y^2 = 25$ have in common exactly ?

(A) 0 points
(B) 1 point
(C) 2 points
(D) 3 points
(E) 4 points

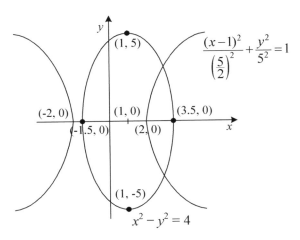

SAT II Math – Formula and Notes

4) Parabola: Let P_4 be the plane that cuts the cone <u>parallel to slant edge direction</u>.

Then, it creates a parabolic figure with 1 curve, formed by its intersection.

Given $4p(y - k) = (x - h)^2$, or $4p(x - h) = (y - k)^2$,

* vertex = (h, k)
* foci = $(o + h, p + k)$ or $(p + h, o + k)$
* Directrix line: $y = -p + k$ or $x = -p + h$

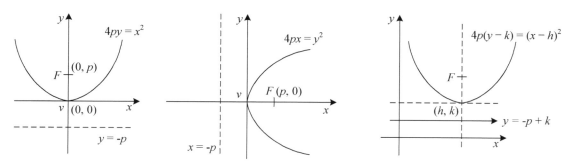

(eg. 1) The focus of a parabola is the point (2,0) and its directrix is the line $x = -2$.
Write an equation of the parabola.

Referring to our previous note on "<u>Conic Sections</u>", the equation of parabola is $4py = x^2$ or $4px = y^2$.
Since directrix line is $x = (-)2$, we use $4px = y^2$, where the focus point is on x-axis with $F = (p, 0) = (2, 0)$.
∴ $p = 2$ ∴ $4(2)x = y^2$, which also have directrix $x = -p = -2$.

16. The ratios between a line and area and volume

Length: 1 $\rightarrow$ 2 $\rightarrow$ 3
Area: 1^2 $\rightarrow$ 2^2 $\rightarrow$ 3^2
Volume: 1^3 $\rightarrow$ 2^3 $\rightarrow$ 3^3

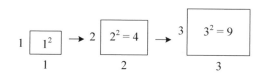

(eg. 1) One side of a given triangle is 6 inches. Inside the triangle a line segment is drawn parallel to this side, cutting off a triangle whose area is one-thirds that of the given triangle.
Find the length of this segment in inches.

Since the ratio of the area $\triangle DBE$ is $1 : \frac{1}{3}$, we get the ratio of the line segment $\overline{AC} : \overline{DE} = 1 : \sqrt{\frac{1}{3}}$.

(Note)   Length: 1 $\rightarrow$ 2 $\rightarrow$ 3
Area: 1^2 $\rightarrow$ 2^2 $\rightarrow$ 3^2
Volume: 1^3 $\rightarrow$ 2^3 $\rightarrow$ 3^3

SAT II Math – Formula and Notes

17. Matrix and Determinant;

- Let $A = \begin{pmatrix} a_{11} & a_{12} \\ a_{21} & a_{22} \end{pmatrix}$ $B = \begin{pmatrix} b_{11} & b_{12} \\ b_{21} & b_{22} \end{pmatrix}$, then, $A \pm B = \begin{pmatrix} a_{11} \pm b_{11} & a_{12} \pm b_{12} \\ a_{21} \pm b_{21} & a_{22} \pm b_{22} \end{pmatrix}$.

  Then, $AB = \begin{pmatrix} a_{11} & a_{12} \\ a_{21} & a_{22} \end{pmatrix}\begin{pmatrix} b_{11} & b_{12} \\ b_{21} & b_{22} \end{pmatrix} = $ (rows of A) $\times$ (columns of B) $= \begin{pmatrix} a_{11}b_{11} + a_{12}b_{21} & a_{11}b_{12} + a_{12}b_{22} \\ a_{21}b_{11} + a_{22}b_{21} & a_{21}b_{12} + a_{22}b_{22} \end{pmatrix}$

- $[\,m \times n\,][\,n \times p\,] = [\,m \times p\,]$, The column of matrix A must be equal to the rows of matrix B!!

- Inverse of A matrix, $A^{-1} = \dfrac{1}{\text{Det } A}\begin{pmatrix} a_{22} & (-)a_{12} \\ (-)a_{21} & a_{11} \end{pmatrix}$

- Det $A = \begin{vmatrix} a_{11} & a_{12} \\ a_{21} & a_{22} \end{vmatrix} = a_{11} \cdot a_{22} - a_{12} \cdot a_{21}$

- If Det $A = 0$, then no inverse matrix of A exists!!!

(eg. 1) If matrix A has dimensions $m \times n$ and matrix B has dimensions $n \times m$, where m and n are distinct positive integers, which of the following statements must be true?

    I.   The product BA does not exist.
    II. The product AB exists and has dimensions $m \times m$.
    III. The product AB exists and has dimensions $n \times n$.

The product of the matrix must match the dimension of columns of the first matrix equal to the rows of the second matrix.

$\therefore [m \times n][n \times p] = [m \times p]$

Here, (I) $BA = [n \times m][m \times n] = [n \times n]$ $\therefore BA$ exist $\therefore$ (I) is not true!

(II) $AB = [m \times n][n \times m] = [m \times m]$ $\therefore AB$ exist with $[m \times m]$

(III) Not true!

18. Variance and Standard Deviation;

- Given data, $x_1, x_2, x_3, \ldots, x_n$, we get Mean $= \bar{x} = \dfrac{x_1 + x_2 + x_3 + \cdots + x_n}{n} = \dfrac{\sum_{i=1}^{n} x_i}{n}$

  Var $= \sigma^2 = \dfrac{[(x_1 - \bar{x})^2 + (x_2 - \bar{x})^2 + \cdots + (x_n - \bar{x})^2]}{n} = \dfrac{\sum_{i=1}^{n}(x_i - \bar{x})^2}{n}$

  Standard Deviation $= \sigma = \sqrt{Variance} = \sqrt{\dfrac{\sum_{i=1}^{n}(x_i - \bar{x})^2}{n}}$

  (Note) The Standard Deviation measures the average of how far the data points are away from the Mean.

(eg. 1) Of the following lists of numbers, which has the biggest standard deviation?

    (A) 2, 5, 8
    (B) 3, 5, 9
    (C) 4, 6, 8
    (D) 1, 9, 18
    (E) 2, 8, 9

In statistics, the standard deviation means "how far each point is away from the mean". But for this problem, without even using the above formula, we intuitively know that the answer choice (D) 1, 9, 18 has the biggest difference between them. Therefore, (D) has the biggest standard deviation.

19. Solid Geometry;

- Volume Sphere: $V = \dfrac{4}{3}\pi r^3$

- Surface Area Sphere: $S = 4\pi r^2$

- Volume Cone: $V = \dfrac{1}{3}\pi r^2 h$

- Lateral Area of Cone: L.A. $= \dfrac{1}{2}\cdot c\cdot l$, where $c = 2\pi r$, $l = \sqrt{r^2 + h^2}$

- Volume Cylinder: $V = \pi r^2 h$

(eg. 1) If the radius of a sphere is doubled, the percent increase in surface area is ?

$V = \dfrac{4}{3}\pi r^3$, $SA = 4\pi r^2$.

Now, r is doubled. $\therefore 2r$

$\therefore V' = \dfrac{4}{3}\pi(2r)^3$, $SA' = 4\pi(2r)^2$

Since $SA' = 16\pi r^2$ vs. $SA = 4\pi r^2$, it is 4:1 ratio.

$\therefore$ the percent difference is 300%.

(eg. 2) If surface area of a sphere has the same numerical value as its volume, what is the length of the radius of this sphere?

The formula for the surface area of sphere is $S = 4\pi r^2$, and the volume $V = \dfrac{4}{3}\pi r^3$.

$\therefore S = V \rightarrow 4\pi r^2 = \dfrac{4}{3}\pi r^3$, or $1 = \dfrac{1}{3}r$

$\therefore r = 3$

SAT II Math – Formula and Notes

(eg. 3)   The radius of a sphere is equal to the radius of the base of the cone. The height of a cone is equal to the radius of its base. The ratio of the volume of the *cone* to the volume of the *sphere* is ?

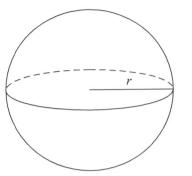
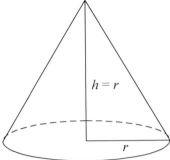

The volume of sphere is $V_1 = \frac{4}{3}\pi r^3$, and the volume of cone is $V_2 = \frac{1}{3}\pi r^2 h = \frac{1}{3}\pi r^2(r) = \frac{1}{3}\pi r^3$

$$\therefore \frac{V_2}{V_1} = \frac{\frac{1}{3}\pi r^3}{\frac{4}{3}\pi r^3} = \frac{1}{4}$$

20. Arc length and area of pie sector;

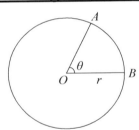

$$\overset{\frown}{AB} = 2\pi r \times \frac{\theta}{360°} = 2\pi r \times \frac{\theta}{2\pi} = r \cdot \theta \text{ in radian mode.}$$

$$\text{Area } OAB = \pi r^2 \times \frac{\theta}{360°} = \pi r^2 \times \frac{\theta}{2\pi} = \frac{1}{2} r^2\theta \text{ in radian mode.}$$

(eg. 1)   A sector of a circle, AQB, with a central angle of $\frac{\pi}{3}$ and a radius of 6 is bent to form a cone with vertex at Q. What is the volume of the cone that is formed?

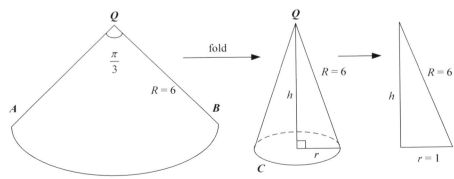

Figure (1)   Figure (2)   Figure (3)

SAT II Math – Formula and Notes

To get the volume of the cone in the figure (2), we need to get the radius r and the height h, because $V = \frac{1}{3}\pi r^2 h$. Notice that the arc length $\widehat{AB}$ becomes the circumference C of the cone. But $\widehat{AB} = R \times \theta = 6 \times \frac{\pi}{3} = 2\pi$, and the formula for the circumference $C = 2\pi r$.

Since $\widehat{AB} = C$, or $2\pi = 2\pi r$, we get $r = 1$. Now, referring to figure (3), we get $h = \sqrt{6^2 - 1^2} = \sqrt{35}$.

Therefore, the volume of the cone is $V = \frac{1}{3}\pi r^2 h$ or $V = \frac{1}{3}\pi(1)^2 \cdot \sqrt{35} = 6.2$.

(eg. 2) A rod, pivoted at one end, rotates through $\frac{\pi}{4}$ radians.

If the rod is 16 inches long, how many inches does the free end travel?

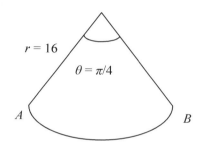

$$\widehat{AB} = r \cdot \theta = 16\,\frac{\pi}{4} = 4\pi$$

21. Polar Coordinates;

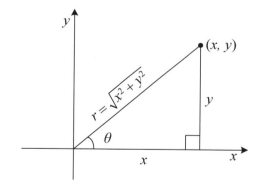

xy-coordinates

polar coordinates

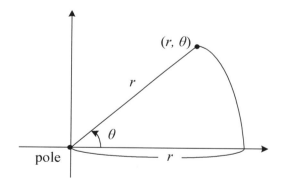

$$r = \sqrt{x^2 + y^2}$$
$$\theta = \tan^{-1}\left(\frac{y}{x}\right)$$

$$x = r\cos\theta$$
$$y = r\sin\theta$$

SAT II Math – Formula and Notes

(eg. 1) If $f(r, \theta) = r \sin \theta$, then $f(3, -1) = ?$

Plugging in $r = 3$, $\theta = -1$ in radian mode, we get $f(3, -1) = 3\sin(-1)$

(eg. 2) The polar coordinates of a point P are $(2, \frac{4\pi}{3})$. The Cartesian (rectangular) coordinates of P are ?

Given $(r, \theta) = (2, \frac{4\pi}{3})$, we have $x = r \cos \theta = 2 \cdot \cos \frac{4\pi}{3} = -1$, and $y = r \sin \theta = 2 \cdot \sin \frac{4\pi}{3} = -\sqrt{3}$

$\therefore (x, y) = (-1, -\sqrt{3})$

(eg. 3) The area of the region enclosed by the graph of the polar curve $r = \dfrac{-2}{\sin \theta - 2\cos \theta}$ and the x- and y-axes is ?

$r = \dfrac{-2}{\sin \theta - 2\cos \theta} \rightarrow r(\sin \theta - 2\cos \theta)$

$= -2 \rightarrow r \sin \theta - 2r \cos \theta = -2$

But $x = r \cos \theta$ and $y = r \sin \theta$.

$\therefore y - 2x = -2$, or $y = 2x - 2$

$\therefore$ area of $\triangle OAB = \dfrac{1}{2}(1)(2) = 1$

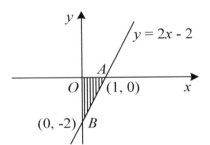

(eg. 4) In Figure 2, $r \cos \theta + r \sin \theta = ?$

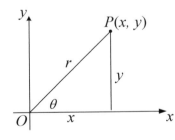

Figure 2

In polar coordinates, we get $x = r \cos \theta$, $y = r \sin \theta$, where $r = \sqrt{x^2 + y^2}$, and $\theta = \tan^{-1} \dfrac{y}{x}$.

$\therefore r \cos \theta + r \sin \theta = x + y$

Model Test No. 01

50 Questions / 60 Minutes

Directions: For each question, determine which of the answer choices is correct and fill in the oval on the answer sheet that corresponds to your choice.

Notes:

1. You will need to use a scientific or graphing calculator to answer some of the questions.
2. Be sure your calculator is in degree mode.
3. Each figure on this test is drawn as accurately as possible unless it is specifically indicated that the figure has not been drawn to scale.
4. The domain of any function f is the set of all real numbers x for which $f(x)$ is also a real number, unless the question indicates that the domain has been restricted in some way.
5. The box below contains five formulas that you may need to answer one or more of the questions.

REFERENCE INFORMATION

THE FOLLOWING INFORMATION IS FOR YOUR REFERENCE IN ANSWERING SOME OF THE QUESTIONS IN THIS TEST.

Volume of a right circular cone with radius r and height h: $V = \dfrac{1}{3}\pi r^2 h$

Lateral Area of a right circular cone with circumference of the base c and slant height l: $S = \dfrac{1}{2}cl$

Volume of a sphere with radius r: $V = \dfrac{4}{3}\pi r^3$

Surface Area of a sphere with radius r: $S = 4\pi r^2$

Volume of a pyramid with base area B and height h: $V = \dfrac{1}{3}Bh$

Answer Sheet
Model Test No. 01

1	Ⓐ Ⓑ Ⓒ Ⓓ Ⓔ	14	Ⓐ Ⓑ Ⓒ Ⓓ Ⓔ	27	Ⓐ Ⓑ Ⓒ Ⓓ Ⓔ	40	Ⓐ Ⓑ Ⓒ Ⓓ Ⓔ
2	Ⓐ Ⓑ Ⓒ Ⓓ Ⓔ	15	Ⓐ Ⓑ Ⓒ Ⓓ Ⓔ	28	Ⓐ Ⓑ Ⓒ Ⓓ Ⓔ	41	Ⓐ Ⓑ Ⓒ Ⓓ Ⓔ
3	Ⓐ Ⓑ Ⓒ Ⓓ Ⓔ	16	Ⓐ Ⓑ Ⓒ Ⓓ Ⓔ	29	Ⓐ Ⓑ Ⓒ Ⓓ Ⓔ	42	Ⓐ Ⓑ Ⓒ Ⓓ Ⓔ
4	Ⓐ Ⓑ Ⓒ Ⓓ Ⓔ	17	Ⓐ Ⓑ Ⓒ Ⓓ Ⓔ	30	Ⓐ Ⓑ Ⓒ Ⓓ Ⓔ	43	Ⓐ Ⓑ Ⓒ Ⓓ Ⓔ
5	Ⓐ Ⓑ Ⓒ Ⓓ Ⓔ	18	Ⓐ Ⓑ Ⓒ Ⓓ Ⓔ	31	Ⓐ Ⓑ Ⓒ Ⓓ Ⓔ	44	Ⓐ Ⓑ Ⓒ Ⓓ Ⓔ
6	Ⓐ Ⓑ Ⓒ Ⓓ Ⓔ	19	Ⓐ Ⓑ Ⓒ Ⓓ Ⓔ	32	Ⓐ Ⓑ Ⓒ Ⓓ Ⓔ	45	Ⓐ Ⓑ Ⓒ Ⓓ Ⓔ
7	Ⓐ Ⓑ Ⓒ Ⓓ Ⓔ	20	Ⓐ Ⓑ Ⓒ Ⓓ Ⓔ	33	Ⓐ Ⓑ Ⓒ Ⓓ Ⓔ	46	Ⓐ Ⓑ Ⓒ Ⓓ Ⓔ
8	Ⓐ Ⓑ Ⓒ Ⓓ Ⓔ	21	Ⓐ Ⓑ Ⓒ Ⓓ Ⓔ	34	Ⓐ Ⓑ Ⓒ Ⓓ Ⓔ	47	Ⓐ Ⓑ Ⓒ Ⓓ Ⓔ
9	Ⓐ Ⓑ Ⓒ Ⓓ Ⓔ	22	Ⓐ Ⓑ Ⓒ Ⓓ Ⓔ	35	Ⓐ Ⓑ Ⓒ Ⓓ Ⓔ	48	Ⓐ Ⓑ Ⓒ Ⓓ Ⓔ
10	Ⓐ Ⓑ Ⓒ Ⓓ Ⓔ	23	Ⓐ Ⓑ Ⓒ Ⓓ Ⓔ	36	Ⓐ Ⓑ Ⓒ Ⓓ Ⓔ	49	Ⓐ Ⓑ Ⓒ Ⓓ Ⓔ
11	Ⓐ Ⓑ Ⓒ Ⓓ Ⓔ	24	Ⓐ Ⓑ Ⓒ Ⓓ Ⓔ	37	Ⓐ Ⓑ Ⓒ Ⓓ Ⓔ	50	Ⓐ Ⓑ Ⓒ Ⓓ Ⓔ
12	Ⓐ Ⓑ Ⓒ Ⓓ Ⓔ	25	Ⓐ Ⓑ Ⓒ Ⓓ Ⓔ	38	Ⓐ Ⓑ Ⓒ Ⓓ Ⓔ		
13	Ⓐ Ⓑ Ⓒ Ⓓ Ⓔ	26	Ⓐ Ⓑ Ⓒ Ⓓ Ⓔ	39	Ⓐ Ⓑ Ⓒ Ⓓ Ⓔ		

USE THIS SPACE FOR SCRATCH WORK

1. $x + \dfrac{1}{x} = 2$, then $x^2 + \dfrac{1}{x^2} =$

   (A) $\dfrac{1}{2}$
   (B) 0
   (C) 2
   (D) 4
   (E) -2

$-\dfrac{1}{2} + \dfrac{1}{-\frac{1}{2}}$

2. How many integers are there in the solution set of $|x - 2| \le 4$?

   (A) 11
   (B) 0
   (C) an infinite number
   (D) 9
   (E) 7

$6 - 2 = 4$
$5 - 2 = 3$
$4 - 2 = 4$
$3 - 2 = 1$
$2 - 2 = 0$

$1 - 2 = -1$
$0 - 2 = -2$
$-1 - 2 = -3$
$-2 - 2 = 4$

3. If $5x - 2y + 4 = 0$ and $y - x^2 = 0$ for $x \ge 0$, then $x =$

   (A) 1.27
   (B) 2.07
   (C) 2.77
   (D) 3.14
   (E) 5.53

$-2y = -5x - 4$

$-2y = -5x - 4$
$y = -2.5x - 2$
$y = x^2$

4. If a point (x, y) is in the 4th quadrant, which of the following must be true?

   I. $x > y$
   II. $x + y > 0$
   III. $\dfrac{x}{y} < 0$

   (A) only I
   (B) only II
   (C) only III
   (D) only I and II
   (E) only I and III

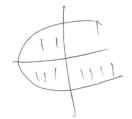

$x = +$
$y = -$

5. Which ordered number pair represents the center of the circle $x^2 + y^2 + 12x - 8y = 0$?

   (A) (9,4)
   (B) (-3,2)
   (C) (3,-2)
   (D) (-6,4)
   (E) (E) (6,4)

35

6. If $f(x) = x^2 + 1$ and $g(x) = 1 + \dfrac{1}{x}$, write the expression $g[f(x)]$ in terms of x.

(A) $1 - \dfrac{1}{x^2}$

(B) $\dfrac{x}{x^2 - 1}$

(C) $\dfrac{x^2 + 2}{x^2 + 1}$

(D) $\dfrac{x^2 + 1}{x^2 + 2}$

(E) none of these

7. The number of terms in the expansion of $(4\sqrt{x} - y^{\frac{2}{3}})^6$ is

(A) 6
(B) 8
(C) 1
(D) 7
(E) 9

8. If $8^x = 4$ and $9^{x+y} = 27$, then $y =$

(A) 2
(B) 5
(C) $\dfrac{25}{2}$
(D) $\dfrac{5}{6}$
(E) 1

9. For what values of k does the graph of
$$\dfrac{(x + k)^2}{3} + \dfrac{(y - 2k)^2}{6} = 1$$ pass through the origin?

(A) only 0
(B) only 1
(C) ± 1
(D) $\pm\sqrt{5}$
(E) no value

10. Suppose i represent and imaginary number, defined by $i = \sqrt{-1}$.
Then which of the following is $i^9 + i^{10} + i^{11} + i^{12} =$

(A) 1
(B) $2i$
(C) $1 - i$
(D) 0
(E) $2 + 2i$

11. The graph of the equation $y = 2x^3 + 3x - 7$

    (A) does not intersect the x-axis
    (B) intersects the x-axis at one and only one point
    (C) intersects the x-axis at exactly three points
    (D) intersects the x-axis at more than three points
    (E) intersects the x-axis at exactly two points

12. The mean weight of the 15 members of an English class was 120 pounds. When a new student enrolled, the mean increased to 121 pounds. What was the weight, in pounds, of the new student?

    (A) 135
    (B) 136
    (C) 137
    (D) 138
    (E) 139

13. Which of the following is *not* an even function?

    (A) $y = |x| + 1$
    (B) $y = \cos x$
    (C) $y = \log x^2$
    (D) $y = x^2 + \tan x$
    (E) $y = 4x^4 + 2x^2 + 3$

14. What is the inverse of the function of
$f(x) = \dfrac{1}{\sqrt{x-1}}, x > 1, y > 0$?

    (A) $\dfrac{1+x^2}{x^2}$ such that $y > 1, x > 0$

    (B) $\dfrac{1-x}{x}, y > 0, x > 1$

    (C) $\dfrac{x^2-1}{x^2}, y > 1, x > 1$

    (D) $\dfrac{1-x^2}{x^2}$ such that $x > -1, y > 0$

    (E) none of these

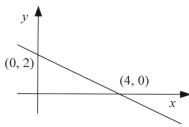

15. A linear function has an *x*-intercept of 4 and a *y*-intercept of 2. The graph of the function has a slope of

(A) 0.77
(B) -0.5
(C) 2.24
(D) 1.29
(E) -0.77

16. If an initial investment of $2,000 in a savings account is compounded continuously at an annual rate of 6 percent, how many years will it take the investment to be worth approximately $5,000?

(A) 4.1
(B) 5.0
(C) 15.3
(D) 20.1
(E) 23.0

17. $f(x) = x^2 - 4x - 7$, find $\lim\limits_{h \to 0} \dfrac{f(x+h) - f(x)}{h}$.

(A) $2x - 4$
(B) 0
(C) ∞
(D) indeterminate
(E) 5

18. The graph of $(x^2 - 1)y = x^2$ has

(A) one horizontal and one vertical asymptote
(B) two vertical but no horizontal asymptotes
(C) one horizontal and two vertical asymptotes
(D) two horizontal and two vertical asymptotes
(E) neither a horizontal nor a vertical asymptote

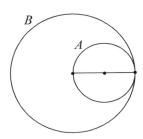

19. If circle A, of area 3 square inches, passes through the center of and is tangent to circle B, then the area of circle B, in square inches, is

(A) 8
(B) $8\sqrt{2}$
(C) $16\sqrt{2}$
(D) 12
(E) 16

20. The remainder obtained when $4x^4 + 3x^3 + 2x^2 - x - 1$ is divided by $x + 1$ is

(A) 5
(B) 0
(C) -3
(D) 3
(E) 13

21. The intersection of a right circular cone with a plane could be which of the following?

I.   A circle
II. An ellipse
III. A triangle

(A) I only
(B) II only
(C) III only
(D) I and III only
(E) I, II, and III

22. Suppose the graph of $f(x) = x^2 + x + 1$ is translated 2 units up and 1 units right. If the resulting graph represents the graph of $g(x)$, what is the value of $g(-1.5)$?

(A) 2.75
(B) 6.75
(C) -0.25
(D) -1.75
(E) 37.25

23. The equation $\dfrac{1}{\sec^2 x} = 2 - \dfrac{1}{\csc^2 x}$ is satisfied by

    (A) all values of x
    (B) exactly two values of x
    (C) only one value of x
    (D) no value of x
    (E) infinitely many but not all values of x

24. If point $P(k, 2k)$ is 3 units from line $3x + 4y = 7$, then k equals

    (A) only 2.75
    (B) only -1.54
    (C) -0.73 or 2
    (D) only 1.54
    (E) -2.75 or 1.54

25. The inverse of $p \rightarrow \sim q$ is equivalent to

    (A) $p \rightarrow \sim q$
    (B) $p \rightarrow q$
    (C) $\sim q \rightarrow p$
    (D) $q \rightarrow p$
    (E) $\sim p \rightarrow \sim q$

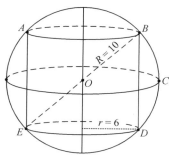

26. A cylinder whose base radius is 6 is inscribed in a sphere of radius 10. What is the difference between the surface area of the sphere and the surface area of the cylinder?

    (A) 354
    (B) 297
    (C) 88
    (D) 427
    (E) 1345

27. If a regular octagonal prism has x pairs of parallel edges, then x equals

    (A) 1
    (B) 2
    (C) 4
    (D) 8
    (E) 12

28. If T varies inversely as the square of S, what is the effect on S when T is doubled?

(A) It is divided by 2.
(B) It is multiplied by 2.
(C) It is multiplied by $\sqrt{2}$.
(D) It is divided by$\sqrt{2}$.
(E) None of the above effects occurs.

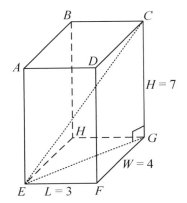

29. A rectangular box has dimensions of length = 3, width = 4, and height = 7. The angle formed by a diagonal of the box with the diagonal of the base of the box contains

(A) 27°
(B) 35°
(C) 40°
(D) 44°
(E) 54°

30. In the equation $x^2 + kx + 48 = 0$, one root is twice the other root. The value(s) of k is (are)

(A) ±6
(B) 14
(C) 6
(D) ±6$\sqrt{6}$
(E) 22

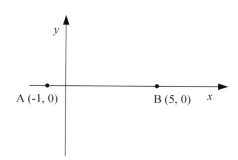

A (-1, 0)   B (5, 0)   x

31. Which of the following is an equation whose graph is the set of points equidistant from the points (-1, 0) and (5, 0)?

    (A) $x = 2$
    (B) $y = 2$
    (C) $x = 2y$
    (D) $y = 2x$
    (E) $y = x + 2$

32. What is the period of the graph of the function $y = \cos^2 x - \sin^2 x$?

    (A) $\dfrac{\pi}{3}$

    (B) $\dfrac{\pi}{2}$

    (C) π

    (D) $\dfrac{3\pi}{2}$

    (E) 2π

33. Two different integers are selected at random from the integers 1 to 10 inclusive. What is the probability that the sum of the two numbers is odd?

    (A) $\dfrac{1}{18}$

    (B) $\dfrac{1}{2}$

    (C) $\dfrac{5}{9}$

    (D) $\dfrac{4}{11}$

    (E) $\dfrac{5}{11}$

34. A sector of a circle, AQB, with a central angle of $\frac{\pi}{3}$ and a radius of 6 is bent to form a cone with vertex at Q. What is the volume of the cone that is formed?

   (A) 8.17
   (B) 6.20
   (C) 4.97
   (D) 5.13
   (E) 12.31

35. Write $[\sqrt{2}(\cos 15° + i \sin 15°)]^3$ in the form $a + bi$.

   (A) $1 + \sqrt{2}i$
   (B) $1 - \sqrt{2}i$
   (C) $\sqrt{2} + i$
   (D) $\sqrt{2} - i$
   (E) $2 + 2i$

36. The graph of the curve whose parametric equations are $x = \dfrac{a}{\cos \theta}$ and $y = b \tan \theta$ is a(n):

   (A) ellipse
   (B) circle
   (C) parabola
   (D) hyperbola
   (E) straight line

37. If $f(x) = x^2 - 2x + 4$, what must the value of k be equal to, so that the graph of $f(x + k)$ will be symmetric to the y-axis?

   (A) 1
   (B) $-\dfrac{1}{2}$
   (C) 0
   (D) $-\dfrac{1}{4}$
   (E) $\dfrac{1}{4}$

38. If $\log_r 5 = A$ and $\log_r 15 = B$, then $\log_r \dfrac{r}{9}$ is equal to

   (A) $1 - A + B$
   (B) $1 + 2A - 2B$
   (C) $1 + A - B$
   (D) $1 - 2A + 2B$
   (E) zero, if $r = 4$

39. $\tan (\text{arc sin } \frac{3}{5})$ equals

    (A) 0.95
    (B) 0.75
    (C) 0.33
    (D) 0.35
    (E) 0.50

40. If a geometric sequence begins with the terms $\frac{1}{4}$, 2, ...,
    what is the sum of the first 7 terms?

    (A) $74898 \frac{1}{4}$

    (B) 56561

    (C) $9362 \frac{1}{4}$

    (D) $1170 \frac{1}{4}$

    (E) 6

41. If $\arccos (\sin x) = \frac{\pi}{6}$ and $0 \leq x \leq \frac{\pi}{2}$, then x could equal

    (A) 0

    (B) $\frac{\pi}{6}$

    (C) $\frac{\pi}{4}$

    (D) $\frac{\pi}{3}$

    (E) $\frac{\pi}{2}$

42. If vector $\vec{v} = (\sqrt{2}, \sqrt{3})$ and vector $\vec{u} = (1, -1)$,
    find the value of $|\sqrt{3}\,\vec{v} + 3\,\vec{u}\,|$.

    (A) 52
    (B) $3 + \sqrt{6}$
    (C) 6
    (D) $3 - \sqrt{6}$
    (E) 3

43. If $\tan A = \cot B$, which of the following must be true?

    (A) $A = B$
    (B) $A + B = 0$
    (C) $A + B = 90°$
    (D) $A + B = 180°$
    (E) $A - B = 180°$

44

44. If the mean of the set of data 2, 3, 3, 7, 1, 5, x is $4.\overline{271}$, what is the value of x?

(A) 8.9
(B) -10.7
(C) 5.6
(D) 2.5
(E) 7.4

45. A committee of 4 people is to be selected from 5 men and 7 women. If the selection is made randomly, what is the probability that the committee consists of 2 men and 2 women?

(A) $\dfrac{1}{3}$

(B) $\dfrac{14}{33}$

(C) $\dfrac{35}{144}$

(D) $\dfrac{4}{35}$

(E) $\dfrac{35}{495}$

46. The graph of $y = \log_2 \dfrac{1}{x}$ and $y = \ln \dfrac{x^2}{3}$ intersect at a point where x equals

(A) 6.24
(B) 1.38
(C) 1.69
(D) 1.05
(E) 5.44

47. Which of the following is equivalent to $\sin\left(\theta - \dfrac{\pi}{3}\right) + \cos\left(\theta - \dfrac{\pi}{6}\right)$ for all values of θ?

(A) $\sin\theta$
(B) $\cos\theta$
(C) $\sqrt{3}\sin\theta + \cos\theta$
(D) $\sqrt{3}\sin\theta$
(E) $\sqrt{3}\cos\theta$

48. What is the probability that a prime number is less than 11, given that it is less than 19?

(A) $\dfrac{1}{2}$

(B) $\dfrac{3}{4}$

(C) $\dfrac{4}{7}$

(D) $\dfrac{5}{8}$

(E) $\dfrac{2}{3}$

49. If $(\csc x)(\cot x) < 0$, which of the following must be true?

I. $\tan x < 0$
II. $(\sin x)(\cos x) < 0$
III. x is in the second or third quadrant

(A) I only
(B) II only
(C) III only
(D) II and III
(E) I and II

50. If the graph below represents the function $f(x)$, which of the following could represent the equation of the inverse of f?

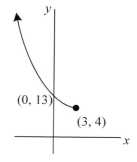

(A) $x = y^2 - 7y - 2$
(B) $x = y^2 + 2$
(C) $x = (y - 3)^2 + 4$
(D) $x = (y + 2)^2 - 3$
(E) $x = (y + 4)^2 + 3$

Answer Key
Model Test No. 01

#	Ans	#	Ans	#	Ans	#	Ans
1	C	14	A	27	E	40	A
2	D	15	B	28	D	41	D
3	D	16	C	29	E	42	B
4	E	17	A	30	D	43	C
5	D	18	C	31	A	44	A
6	C	19	D	32	C	45	B
7	D	20	D	33	C	46	B
8	D	21	E	34	B	47	A
9	C	22	B	35	E	48	C
10	D	23	D	36	D	49	C
11	B	24	C	37	A	50	C
12	B	25	C	38	B		
13	D	26	D	39	B		

How to Score the SAT Subject Test in Mathematics Level 2

When you take an actual SAT Subject Test in Mathematics Level 2, your answer sheet will be "read" by a scanning machine that will record your responses to each question. Then a computer will compare your answers with the correct answers and produce your raw score. You get one point for each correct answer. For each wrong answer, you lose one-fourth of a point. Questions you omit (and any for which you mark more than one answer) are not counted. This raw score is converted to a scaled score that is reported to you and to the colleges you specify.

Finding Your Raw Test Score

STEP 1: Table A lists the correct answers for all the questions on the Subject Test in Mathematics Level 2 that is reproduced in this book. It also serves as a worksheet for you to calculate your raw score.

- Compare your answers with those given in the table.
- Put a check in the column marked "Right" if your answer is correct.
- Put a check in the column marked "Wrong" if your answer is incorrect.
- Leave both columns blank if you omitted the question.

STEP 2: Count the number of right answers.
Enter the total here: _____

STEP 3: Count the number of wrong answers.
Enter the total here: _____

STEP 4: Multiply the number of wrong answers by .250.
Enter the product here: _____

STEP 5: Subtract the result obtained in Step 4 from the total you obtained in Step 2.
Enter the result here: _____

STEP 6: Round the number obtained in Step 5 to the nearest whole number.
Enter the result here: _____

The number you obtained in Step 6 is your raw score.

Scaled Score Conversion Table
Subject Test in Mathematics Level 2

Raw Score	Scaled Score	Raw Score	Scaled Score	Raw Score	Scaled Score
50	800	28	630	6	470
49	800	27	630	5	460
48	800	26	620	4	450
47	800	25	610	3	440
46	800	24	600	2	430
45	800	23	600	1	420
44	800	22	590	0	410
43	790	21	580	-1	400
42	780	20	580	-2	390
41	770	19	570	-3	370
40	760	18	560	-4	360
39	750	17	560	-5	350
38	740	16	550	-6	340
37	730	15	540	-7	340
36	710	14	530	-8	330
35	700	13	530	-9	330
34	690	12	520	-10	320
33	680	11	510	-11	310
32	670	10	500	-12	300
31	660	9	490		
30	650	8	480		
29	640	7	480		

1. $x + \dfrac{1}{x} = 2$, then $x^2 + \dfrac{1}{x^2} =$

    (A) $-\dfrac{1}{2}$ (B) 0   (C) 2   (D) 4   (E) -2

Since $x + \dfrac{1}{x} = 2$, by squaring both side of the equation,

we get $x^2 + 2x\left(\dfrac{1}{x}\right) + \left(\dfrac{1}{x}\right)^2 = 4$, or simplifying it,

$x^2 + 2 + \dfrac{1}{x^2} = 4$. Therefore, $x^2 + \dfrac{1}{x^2} = 4 - 2 = 2$.

Ans. (C)

---

2. How many integers are there in the solution set of $|x - 2| \le 4$?

    (A) 11
    (B) 0
    (C) an infinite number
    (D) 9
    (E) 7

Here, $|x - 2| \le 4$ really means that the absolute value of $(x - 2)$ must be less than 4. This implies that the quantity of $(x - 2)$ runs from $(-)4$ to $(+)4$, or $-4 \le (x - 2) \le 4$.
By adding 2, we get $-4 + 2 \le x \le 4 + 2$ or $-2 \le x \le 6$.
Therefore, there are 9 integers in the interval.

Ans. (D)

---

3. If $5x - 2y + 4 = 0$ and $y - x^2 = 0$ for $x \ge 0$, then $x =$

    (A) 1.27
    (B) 2.07
    (C) 2.77
    (D) 3.14
    (E) 5.53

$5x - 2y + 4 = 0$ -- ①
$y - x^2 = 0$ -- ②
From eq. ②, we get $y = x^2$.
Now, substituting $y = x^2$ into eq. ①,
we get $5x - 2x^2 + 4 = 0$, or $2x^2 - 5x - 4 = 0$.
Solving the above eq., we may use quadratic formula,
$x = \dfrac{-b \pm \sqrt{b^2 - 4ac}}{2a}$, where $a = 2$, $b = -5$, $c = -4$.
The only positive solution is, $x = \dfrac{5 + \sqrt{57}}{4} = 3.14$.

Ans. (D)

---

4. If a point (x, y) is in the 4th quadrant, which of the following must be true?

    I. $x > y$
    II. $x + y > 0$
    III. $\dfrac{x}{y} < 0$

    (A) only I
    (B) only II
    (C) only III
    (D) only I and II
    (E) only I and III

Since (x, y) is in the 4th quadrant, we get $x > 0$ and $y < 0$.
Therefore,
I. $x > y$ is a true statement
II. $x + y > 0$ may or may not be true,
     dependry on x, y values.
III. $\dfrac{x}{y} < 0$ is true, because x and y have opposite signs.

Ans. (E)

---

5. Which ordered number pair represents the center of the circle $x^2 + y^2 + 12x - 8y = 0$?

    (A) (9,4)   (B) (-3,2)   (C) (3,-2)
    (D) (-6,4)   (E) (6,4)

To get the center of this circle,
we need to modify this eq. into a <u>complete square</u> form.
That is, $x^2 + 12x + y^2 - 8y = 0$ becomes $(x + 6)^2 + (y - 4)^2 = 52$.
Therefore, we get the center $(-6, 4)$.

Ans. (D)

6. If $f(x) = x^2 + 1$ and $g(x) = 1 + \dfrac{1}{x}$, write the expression $g[f(x)]$ in terms of x.

(A) $1 - \dfrac{1}{x^2}$ (B) $\dfrac{x}{x^2 - 1}$

(C) $\dfrac{x^2 + 2}{x^2 + 1}$ (D) $\dfrac{x^2 + 1}{x^2 + 2}$

(E) none of these

$g(f(x))$ is what we call a <u>composite function</u>, which means, it is a function of a function. To solve, we need to get $g(x)$ first, and then replace x with $f(x)$.

That is, $g(f(x)) = 1 + \dfrac{1}{f(x)} = 1 + \dfrac{1}{x^2 + 1} = \dfrac{x^2 + 2}{x^2 + 1}$

Ans. (C)

---

7. The number of terms in the expansion of $(4\sqrt{x} - y^{\frac{2}{3}})^6$ is

(A) 6
(B) 8
(C) 1
(D) 7
(E) 9

This is a Binomial Expansion. Refer to the formula,
$(A + B)^r = \sum_{n=1}^{r} {}_nC_r A^{n-r} B^r = {}_nC_0 A^n B^0 + {}_nC_1 A^{n-1} B^1 + {}_nC_2 A^{n-2} B^2 + \cdots + {}_nC_n A^0 B^n$,
which has $(n + 1)$ term, for the power of n-expansion.
Since our $n = 6$, we get $(6 + 1)$ term.

Ans. (D)

---

8. If $8^x = 4$ and $9^{x+y} = 27$, then $y =$

(A) 2

(B) 5

(C) $\dfrac{25}{2}$

(D) $\dfrac{5}{6}$

(E) 1

To solve this exponential equation, we need to convert every base into the same base number.
That is, $8^x = (2^3)^4 = 2^{3x}$ -- ①,
$4 = 2^2$ -- ②.

Equating ① and ②, we get $2^{3x} = 2^2$, thus, $3x = 2$ or $x = \dfrac{2}{3}$.

For the second equation,
$9^{x+y} = 27$ becomes $(3^2)^{x+y} = 3^3$, or $2(x + y) = 3$.

$2(\dfrac{2}{3} + y) = 3$ or $y = \dfrac{3}{2} - \dfrac{2}{3} = \dfrac{5}{6}$.

Ans. (D)

---

9. For what values of k does the graph of $\dfrac{(x+k)^2}{3} + \dfrac{(y-2k)^2}{6} = 1$ pass through the origin?

(A) only 0
(B) only 1
(C) ±1
(D) ±√5
(E) no value

The expression of "passing through the origin" means that we need to substitute $x = 0$, and $y = 0$ into the equation.

Therefore, we get $\dfrac{(0+k)^2}{3} + \dfrac{(0-2k)^2}{6} = 1$,

or $\dfrac{k^2}{3} + \dfrac{4k^2}{6} = \dfrac{2k^2}{6} + \dfrac{4k^2}{6} = \dfrac{6k^2}{6} = k^2 = 1$.
Thus, $k = ±1$.

Ans. (C)

---

10. Suppose i represent and imaginary number, defined by $i = \sqrt{-1}$. Then which of the following is $i^9 + i^{10} + i^{11} + i^{12} =$

(A) 1
(B) $2i$
(C) $1 - i$
(D) 0
(E) $2 + 2i$

Since $i = \sqrt{-1}$, we get $i^2 = (-)1$, $i^3 = i^2 \times i = (-1)i = (-)i$, $i^4 = (i^2)(i^2) = (-1)(-1) = 1$.
Please note that i^4 becomes one, which means i^n has a cycle of four, and $i^4 = 1$, $i^8 = 1$, $i^{12} = 1$, ... so on.
Therefore, $i^9 = i^1$, because $9 \div 4 = 2$ with its remainder of 1.
So is, $i^{10} = i^2$, because $(10 \div 4)$ gives its remainder of 2.
$i^{11} = i^3$ and $i^{12} = 1$, because $(12 \div 4)$ gives 0 remainder.
Therefore, $i^9 + i^{10} + i^{11} + i^{12} = i^1 + i^2 + i^3 + i^0 = i + (-1) + (-i) + 1 = 0$.

Ans. (D)

11. The graph of the equation $y = 2x^3 + 3x - 7$

    (A) does not intersect the x-axis
    (B) intersects the x-axis at one and only one point
    (C) intersects the x-axis at exactly three points
    (D) intersects the x-axis at more than three points
    (E) intersects the x-axis at exactly two points

For this problem, we will use a graphing calculator by putting $y_1 = 2x^3 + 3x - 7$ in graphing utility.

Ans. (B)

12. The mean weight of the 15 members of an English class was 120 pounds. When a new student enrolled, the mean increased to 121 pounds.
What was the weight, in pounds, of the new student?

    (A) 135
    (B) 136
    (C) 137
    (D) 138
    (E) 139

Since the mean weight of 15 members was 120 pounds, we get the sum of all the weights of 15 members as,
sum $S = 120 \times 15 = 1800$.
Now, let x = weight of a new student,
then the mean equals, $\dfrac{1800+x}{15+1} = 121$.
Solving this eq. we get $x = 136$.

Ans. (B)

13. Which of the following is *not* an even function?

    (A) $y = |x| + 1$
    (B) $y = \cos x$
    (C) $y = \log x^2$
    (D) $y = x^2 + \tan x$
    (E) $y = 4x^4 + 2x^2 + 3$

The definition of even function is, $f(-x) = f(x)$, or symmetric to y-axis. In this problem, instead of using the definition, it would be better to use a graphing calculator to see if any of the function given is not symmetric to y-axis. The choice (D) $y = x^2 + \tan x$ is not symmetric to y-axis.

Ans. (D)

14. What is the inverse of the function of
$f(x) = \dfrac{1}{\sqrt{x-1}}$, $x > 1$, $y > 0$?

    (A) $\dfrac{1+x^2}{x^2}$ such that $y > 1$, $x > 0$

    (B) $\dfrac{1-x}{x}$, $y > 0$, $x > 1$

    (C) $\dfrac{x^2-1}{x^2}$, $y > 1$, $x > 1$

    (D) $\dfrac{1-x^2}{x^2}$ such that $x > -1$, $y > 0$

    (E) none of these

The inverse of a function $f(x)$ is defined by $f^{-1}(x)$, for which,

  1. $f(x)$ must be one to one function.
  2. $f(x)$ and $f^{-1}(x)$ are symmetric to $y = x$ line.
  3. To get $f^{-1}(x)$, we need to switch x into y, and y into x.
  4. The Domain and Range of $f(x)$ become the Range and the Domain of $f^{-1}(x)$ by interchanging respectively.

Here, $f(x) = \dfrac{1}{\sqrt{x-1}} = y$. Now, by interchanging x and y,

we get $\dfrac{1}{\sqrt{y-1}} = x$ or $\left(\dfrac{1}{\sqrt{y-1}}\right)^2 = x^2$.

That is $\dfrac{1}{y-1} = x^2$ or $y - 1 = \dfrac{1}{x^2}$,

which results in $y = \dfrac{1}{x^2} + 1 = \dfrac{1+x^2}{x^2}$ with $x > 0$, $y > 1$.

Ans. (A)

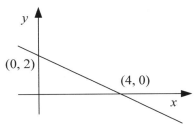

15. A linear function has an *x*-intercept of 4 and a *y*-intercept of 2. The graph of the function has a slope of

 (A) 0.77
 (B) -0.5
 (C) 2.24
 (D) 1.29
 (E) -0.77

16. If an initial investment of $2,000 in a savings account is compounded continuously at an annual rate of 6 percent, how many years will it take the investment to be worth approximately $5,000?

 (A) 4.1
 (B) 5.0
 (C) 15.3
 (D) 20.1
 (E) 23.0

17. $f(x) = x^2 - 4x - 7$, find $\lim\limits_{h \to 0} \dfrac{f(x+h)-f(x)}{h}$.

 (A) $2x - 4$
 (B) 0
 (C) ∞
 (D) indeterminate
 (E) 5

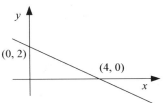

Referring this linear graph, we get the slope, $m = \dfrac{\text{rise}}{\text{run}} = \dfrac{2}{4}$,
but the slope is in negative direction, thus, $m = -\dfrac{2}{4} = -\dfrac{1}{2}$.

Ans. (B)

Notice that the saving is compounded <u>continuously</u> at $r = 6\%$ or, $r = 0.06$ annually. For problems such as calculating interest rate, population growth or any exponential growth, we have the following formula in two different cases.

1. Discrete case: $y = A_0 (1 + \dfrac{r}{n})^{nt}$, where r is annual rate, A_0 is an initial amount, n is the number of terms applied to annual rate, and t is time period.
2. Continuous case: $y = A_0 e^{rt}$.

In this problem, we choose the continuous case, because it is compounded continuously.
$y = A_0 e^{rt}$, where $A_0 = 2000$, $r = 0.06$, $y = 5,000$.
This leads us to our eq. $5000 = 2000 \times e^{0.06t}$, or $\dfrac{5000}{2000} = e^{0.06t}$.

Applying natural log, ln on both sides, we get $\ln \dfrac{5}{2} = 0.06t$.

Therefore, $t = \dfrac{\ln \dfrac{5}{2}}{0.06} = 15.3$

Ans. (C)

This problem involves with the definition of Derivative in Calculus. Of course, if you have learned in Calculus, then, just take the derivative of $f(x)$, or $f'(x) = 2x - 4$ is the answer.
Otherwise, we have to calculate,
$f(x + h) = (x + h)^2 - 4(x + h) - 7$ --- eq.①
$f(x) = x^2 - 4x - 7$ --- eq.②
Thus, $f(x + h) - f(x) = $ eq.① - eq.②
$= (x^2 + 2xh + h^2 - 4x - 4h - 7) - (x^2 - 4x - 7) = 2xh - 4h + h^2$.
Now, dividing by h gives us,
$\lim\limits_{h \to 0} \dfrac{2xh - 4h + h^2}{h} = \lim\limits_{h \to 0} (2x - 4 + h) = 2x - 4$.

Ans. (A)

18. The graph of $(x^2 - 1)y = x^2$ has

(A) one horizontal and one vertical asymptote
(B) two vertical but no horizontal asymptotes
(C) one horizontal and two vertical asymptotes
(D) two horizontal and two vertical asymptotes
(E) neither a horizontal nor a vertical asymptote

By dividing both side of the eq's, we get $y = \dfrac{x^2}{x^2-1}$, where $x \neq (-)1$, or $x \neq 1$. Now, this is a rational function with both horizontal asymptotes, and vertical asymptotes. To get the horizontal asymptotes, we try $\lim\limits_{x\to\infty} \dfrac{x^2}{x^2-1}$, and since the degrees of numerator and denominator are the same, which is 2, the limit value in this case, is the ratio of the leading coefficients $\dfrac{1}{1}$. That is, our horizontal asymptote is $y = \dfrac{1}{1} = 1$. Now, to get the vertical asymptotes, where the function is undefined, we set its denominator equal to zero. That is, $x^2 - 1 = 0$. Our vertical asymptotes are $x = 1$, $x = (-)1$.

(Note: Be careful about removable discontinuity, where the numerator and denominator can be REDUCED by factoring!!)
In this problem, however, we may use a graphing calculator, which can be a much more effective way to solve quickly during the exam. You may find the asymptotes from the graphing of $y = \dfrac{x^2}{x^2-1}$, where we find two vertical asymptotes at $x = 1$, $x = -1$, and one horizontal asymptote at $y = 1$.

Ans. (C)

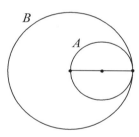

19. If circle A, of area 3 square inches, passes through the center of and is tangent to circle B, then the area of circle B, in square inches, is

(A) 8
(B) $8\sqrt{2}$
(C) $16\sqrt{2}$
(D) 12
(E) 16

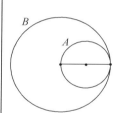

Since area of circle A is 3, we get its radius r from the eq, $3 = \pi r^2$, which leads to $r = \sqrt{\dfrac{3}{\pi}}$.
Now, the diameter D of the circle A becomes the radius R of the bigger circle B, $\therefore R = 2\sqrt{\dfrac{3}{\pi}}$.
Then the area of circle B is, $\pi R^2 = \pi \left(2\sqrt{\dfrac{3}{\pi}}\right)^2 = 12$.

Ans. (D)

20. The remainder obtained when
$4x^4 + 3x^3 + 2x^2 - x - 1$ is divided by $x + 1$ is

(A) 5
(B) 0
(C) -3
(D) 3
(E) 13

To solve this problem, we need to use what we call, "Remainder Theorem". The Remainder Theorem says, when $P(x)$ is divided by $(x - a)$, then the remainder becomes $P(a)$. So in this case, for the given polynomial function , $P(x) = 4x^4 + 3x^3 + 2x^2 - x - 1$, when this function is being divided by $(x + 1)$, we get the remainder, $P(x = -1) = P(-1) = 4(-1)^4 + 3(-1)^3 + 2(-1)^2 - (-1) - 1 = 3$.

Ans. (D)

21. The intersection of a right circular cone with a plane could be which of the following?

    I.   A circle
    II. An ellipse
    III. A triangle

    (A) I only
    (B) II only
    (C) III only
    (D) I and III only
    (E) I, II, and III

In this problem, all of the circle, ellipse and triangle are possible, depending on how it is cut.

    I.   A circle is possible by cutting in horizontal direction.
    II. An ellipse is possible by cutting in a tilted direction.
    III. A triangle is possible by cutting in a vertical direction

Ans. (E)

---

22. Suppose the graph of $f(x) = x^2 + x + 1$ is translated 2 units up and 1 units right. If the resulting graph represents the graph of $g(x)$, what is the value of $g(-1.5)$?

    (A) 2.75
    (B) 6.75
    (C) -0.25
    (D) -1.75
    (E) 37.25

$y = f(x - h) + k$, represents the standard form in which $f(x)$ has a horizontal translating of h unit, and a vertical shifting of k unit.
When $h > 0$, then move to the right, and $h < 0$, move to the left etc…
Now, $g(x)$ represents the graph of $f(x)$ with $h = 1$, and $k = 2$.
Therefore, $g(x) = f(x - 1) + 2$, or plugging in $(x - 1)$ for x, and adding 2, we get $g(x) = [(x - 1)^2 + (x - 1) + 1] + 2$
$= (x - 1)^2 + (x - 1) + 3$.
Thus, $g(-1.5) = [(-2.5)^2 + (-2.5) + 3] = 6.75$.

Ans. (B)

---

23. The equation $\dfrac{1}{\sec^2 x} = 2 - \dfrac{1}{\csc^2 x}$ is satisfied by

    (A) all values of x
    (B) exactly two values of x
    (C) only one value of x
    (D) no value of x
    (E) infinitely many but not all values of x

Refer to $\dfrac{1}{\sin x} = \csc x$. And $\dfrac{1}{\cos x} = \sec x$.
Therefore, by replacing sec x and csc x,
we get $\dfrac{1}{\left(\frac{1}{\cos x}\right)^2} = 2 - \dfrac{1}{\left(\frac{1}{\sin x}\right)^2}$, or $\cos^2 x = 2 - \sin^2 x$,
which is $\sin^2 x + \cos^2 x = 2$.
But, by Identity Rule of Trigonometry,
we have $\sin^2 x + \cos^2 x = 1$.
Our eq, $\sin^2 x + \cos^2 x = 2$ contradicts the above rule.

Ans. (D)

---

24. If point $P(k, 2k)$ is 3 units from line $3x + 4y = 7$, then k equals

    (A) only 2.75
    (B) only -1.54
    (C) -0.73 or 2
    (D) only 1.54
    (E) -2.75 or 1.54

The formula for the distance D between a point (x, y) and a line $Ax + By + C = 0$ is, $D = \dfrac{|Ax_1 + By_1 + C|}{\sqrt{A^2 + B^2}}$.
Using this formula with $(x_1, y_1) = (k, 2k)$ and $A = 3$, $B = 4$,
$C = (-)7$, we get $D = \dfrac{|3(k) + 4(2k) - 7|}{\sqrt{3^2 + 4^2}} = 3$, or $(11k - 7) = \pm15$.
Thus, $11k = 7 \pm 15$, or $k = 2$, $k = (-)0.73$.

Ans. (C)

25. The inverse of $p \rightarrow \sim q$ is equivalent to

(A) $p \rightarrow \sim q$
(B) $p \rightarrow q$
(C) $\sim q \rightarrow p$
(D) $q \rightarrow p$
(E) $\sim p \rightarrow \sim q$

For this problem, we need to understand LOGIC in four different ways. Let's make a statement. If p, then q. Using mathematical symbol for this statement is, $p \rightarrow q$.

I.   Negation: If p, then <u>NOT</u> q. Or $p \rightarrow \sim q$, where "$\sim$" sign represents "NOT".
II. Inverse: If <u>NOT</u> p, then <u>NOT</u> q, or $\sim p \rightarrow \sim q$.
III. Converse: If q, then p. Or $q \rightarrow p$.
IV. Contra Positive: If <u>NOT</u> q, then <u>NOT</u> p. Or $\sim q \rightarrow \sim p$.

Of all these, the Contra Positive is the most important one to remember. Here, the Contra Positive is exactly the same statement as the original statement. Often, if we can not prove the original statement, then we may use the logic of Contra Positive to prove the original statement.
Now, based on the above facts, the inverse of $p \rightarrow \sim q$ is equivalent to negation on both, or $\sim p \rightarrow \sim(\sim q)$, which is $\sim p \rightarrow q$, because $\sim(\sim q) = q$ itself.
However, we don't have any answer choice for this.
Therefore, making Contra Positive of $\sim p \rightarrow q$ by switching $\sim p$ and q, and negating both, we get $\sim q \rightarrow \sim(\sim p)$ or $\sim q \rightarrow p$.

Ans. (C)

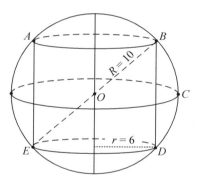

26. A cylinder whose base radius is 6 is inscribed in a sphere of radius 10. What is the difference between the surface area of the sphere and the surface area of the cylinder?

(A) 354
(B) 297
(C) 88
(D) 427
(E) 1345

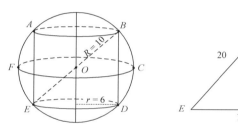

Referring to the picture, since the cylinder is inscribed inside the sphere with radius $R = 10$, we get $\overline{OB} = \overline{OE} = 10$, or $\overline{BE} = 20$. Also, $\overline{ED} = 2 \times$ the radius of the cylinder, or $\overline{ED} = 2 \times r = 2 \times 6 = 12$.
Now, considering $\triangle BED$, we have $\overline{BD} = h = \sqrt{20^2 - 12^2}$ $= 16$, which is the height of the cylinder.
Therefore, the surface area of sphere, S_1, minus the surface area of cylinder, S_2, will be,
$S_1 - S_2 = (4\pi R^2) - (2\pi r^2 + 2\pi rh) =$
$4 \times \pi \times 10^2 - (2\pi \times 6^2 + 2\pi \times 6 \times 16) = 427.26$

Ans. (D)

27. If a regular octagonal prism has x pairs of parallel edges, then x equals

(A) 1
(B) 2
(C) 4
(D) 8
(E) 12

A regular Octagonal Prism has two octagon faces of top and bottom with 8 sides, which forms four different pairs of parallel edges. Thus, a total of $2 \times 4 = 8$ pairs, on top and bottom faces. Also, it has another 4 pairs of parallel edges that shape the column of the prism. These are the parallel edges that connect each the vertices of the top and the bottom faces. Therefore, the sum of all parallel edges is 12 pairs.

Ans. (E)

28. If T varies inversely as the square of S, what is the effect on S when T is doubled?

(A) It is divided by 2.
(B) It is multiplied by 2.
(C) It is multiplied by $\sqrt{2}$.
(D) It is divided by $\sqrt{2}$.
(E) None of the above effects occurs.

When T varies <u>inversely</u> as the square of S, we set the equation as, $T = k \times \dfrac{1}{S^2}$, **where $\dfrac{1}{S^2}$ represents the term, 'inversely as the square of S'.**

Now, to get the effect on S when T is doubled,

we have $2T = 2(k \times \dfrac{1}{S^2}) = k \times \dfrac{1}{\left(\frac{S^2}{2}\right)}$, or $k \times \dfrac{1}{\left(\frac{S}{\sqrt{2}}\right)^2}$.

This shows that S must be divided by $\sqrt{2}$.

Ans. (D)

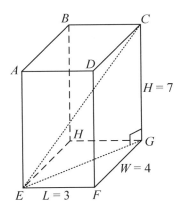

29. A rectangular box has dimensions of length = 3, width = 4, and height = 7. The angle formed by a diagonal of the box with the diagonal of the base of the box contains

(A) 27°
(B) 35°
(C) 40°
(D) 44°
(E) 54°

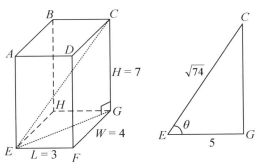

Figure (1)   Figure (2)

Consider the figure (1), given here with dimension of $L = 3$, $W = 4$ and $H = 7$. The diagonal of the box is given by $\overline{EC}$, and $\overline{EC} = \sqrt{3^2 + 4^2 + 7^2} = \sqrt{74}$.

Also, the diagonal of the base is given by $\overline{EG}$, and $\overline{EG} = \sqrt{3^2 + 4^2} = 5$. Now, referring to figure (2),

we have, $\theta = \cos^{-1} \dfrac{5}{\sqrt{74}} = 54$.

Ans. (E)

30. In the equation $x^2 + kx + 48 = 0$, one root is twice the other root. The value(s) of k is (are)

(A) ±6
(B) 14
(C) 6
(D) ±6√6
(E) 22

To solve this problem, we need to refer to the following formula from the quadratic eq.:

Given $ax^2 + bx + c = 0$, suppose its zeros are x_1, and x_2.

Then, the sum of $x_1 + x_2 = -\dfrac{b}{a}$, and the product $x_1 \times x_2 = \dfrac{c}{a}$.

Now, from our equation, $x^2 + kx + 48 = 0$, we get $a = 1$, $b = k$ and $c = 48$. Also, the problem states that one root is twice of the other root, this, $x_2 = 2x_1$.

Now, the sum $x_1 + x_2 = x_1 + 2x_1 = 3x_1 = -\dfrac{b}{a}$,

or $3x_1 = -\dfrac{k}{1} = -k$, or $x_1 = -\dfrac{k}{3}$. --- eq.(1)

Also, the product $x_1 \times x_2 = x_1 \times (2x_1) = 2x_1^2 = \dfrac{c}{a}$,

or $2x_1^2 = \dfrac{48}{1} = 48$, or $x_1^2 = 24$. --- eq.(2)

Therefore, from the above eq.(1) and (2),

we get $x_1 = \pm 2\sqrt{6} = -\dfrac{k}{3}$, and thus, $k = \pm 6\sqrt{6}$.

Ans. (D)

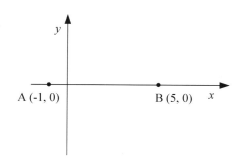

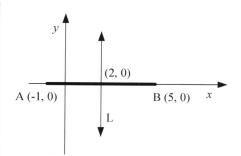

31. Which of the following is an equation whose graph is the set of points equidistant from the points (-1, 0) and (5, 0)?

    (A) $x = 2$
    (B) $y = 2$
    (C) $x = 2y$
    (D) $y = 2x$
    (E) $y = x + 2$

The set of all points that is equidistance from the line segment $\overline{AB}$ is the perpendicular bisector line, through the midpoint (2, 0) of $\overline{AB}$. Therefore, it is the axis of $x = 2$.

Ans. (A)

32. What is the period of the graph of the function $y = \cos^2 x - \sin^2 x$?

    (A) $\dfrac{\pi}{3}$

    (B) $\dfrac{\pi}{2}$

    (C) π

    (D) $\dfrac{3\pi}{2}$

    (E) 2π

To set the period of the graph of $y = \cos^2 x - \sin^2 x$, we need to make our equation to a general form with a single term:
$y = a \cos(b(x - h)) + k$.

Here, $|a|$ = Amplitude, b = # of frequency, $\dfrac{2\pi}{b}$ = period,

h = units of translation horizontally,
k = units of shifting vertically.

Now, recall from the Trig. Course, that $\cos 2x = \cos^2 x - \sin^2 x$. So, our equation becomes, $y = \cos^2 x - \sin^2 x = \cos 2x$, and thus, the period $p = \dfrac{2\pi}{2} = \pi$.

Ans. (C)

33. Two different integers are selected at random from the integers 1 to 10 inclusive. What is the probability that the sum of the two numbers is odd?

    (A) $\dfrac{1}{18}$

    (B) $\dfrac{1}{2}$

    (C) $\dfrac{5}{9}$

    (D) $\dfrac{4}{11}$

    (E) $\dfrac{5}{11}$

The sum [A] + [B] = ODD. The sum to be ODD, we must have a combination of A and B as:

- (case 1) if A = ODD, then B = EVEN.
  Therefore, the possible choice of A is 1, 3, 5, 7, 9, while the possible choice of B = 2, 4, 6, 8, 10. Since we have 5 choices for A and 5 choices for B, the total possible combination is $(5 \times 5) = 25$.
- (case 2) if A = EVEN, than B = ODD.
  With the same reasoning, we have another 25 possible combination.

Now, for the total number of combination for the sum of two different integers, regardless of EVEN and ODD, will be A = 10 possible integers, B = 9 possible integers without the one that we have already picked for A, and that is, $10 \times 9 = 90$.

Therefore, the answer should be $\dfrac{25+25}{90} = \dfrac{50}{90} = \dfrac{5}{9}$.

Ans. (C)

34. A sector of a circle, AQB, with a central angle of $\frac{\pi}{3}$ and a radius of 6 is bent to form a cone with vertex at Q. What is the volume of the cone that is formed?

(A) 8.17
(B) 6.20
(C) 4.97
(D) 5.13
(E) 12.31

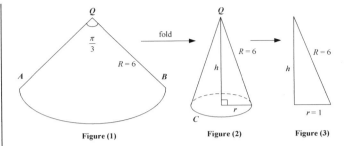

Figure (1)   Figure (2)   Figure (3)

To get the volume of the cone in the figure (2), we need to get the radius r and the height h, because $V = \frac{1}{3}\pi r^2 h$.

Notice that the arc length $\overset{\frown}{AB}$ becomes the circumference C of the cone.

But $\overset{\frown}{AB} = R \times \theta = 6 \times \frac{\pi}{3} = 2\pi$,

and the formula for the circumference $C = 2\pi r$.
Since $\overset{\frown}{AB} = C$, or $2\pi = 2\pi r$, we get $r = 1$.
Now, referring to figure (3), we get $h = \sqrt{6^2 - 1^2} = \sqrt{35}$.
Therefore, the volume of the cone is

$V = \frac{1}{3}\pi r^2 h$ or $V = \frac{1}{3}\pi (1)^2 \cdot \sqrt{35} = 6.2$.

Ans. (B)

35. Write $[\sqrt{2}(\cos 15° + i \sin 15°)]^3$ in the form $a + bi$.

(A) $1 + \sqrt{2}i$
(B) $1 - \sqrt{2}i$
(C) $\sqrt{2} + i$
(D) $\sqrt{2} - i$
(E) $2 + 2i$

Referring to <u>De Moivres Theorem</u> for the Complex Number, Let the complex number, $Z = a + bi = r(\cos\theta + i\sin\theta)$,

where $r = \sqrt{a^2 + b^2}$, $\theta = \tan^{-1}\frac{b}{a}$.
Then, $Z^n = (a + bi)^n = [r(\cos\theta + i\sin\theta)]^n$
$= r^n[\cos(n\theta) + i\sin(n\theta)]$.
Using this formula, we get $[\sqrt{2}(\cos 15° + i\sin 15°)]^3$
$= (\sqrt{2})^3[\cos(3\times15°) + i\sin(3\times15°)]$
$= 2\sqrt{2}(\cos 45° + i\sin 45°) = 2\sqrt{2}(\frac{1}{\sqrt{2}} + i\cdot\frac{1}{\sqrt{2}}) = 2 + 2i$

Ans. (E)

36. The graph of the curve whose parametric equations are $x = \frac{a}{\cos\theta}$ and $y = b\tan\theta$ is a(n):

(A) ellipse
(B) circle
(C) parabola
(D) hyperbola
(E) straight line

To solve this problem, we need to recall the definition of $\sec\theta = \frac{1}{\cos\theta}$, and the identity rule $1 + \tan^2\theta = \sec^2\theta$.

Here, $x = \frac{a}{\cos\theta}$, therefore, $\cos\theta = \frac{x}{a}$, or $\frac{1}{\cos\theta} = \sec\theta = \frac{x}{a}$.

Also, $y = b\tan\theta$ becomes $\tan\theta = \frac{y}{b}$.

This, using the identity rule of $1 + \tan^2\theta = \sec^2\theta$,

we get $1 + \left(\frac{y}{b}\right)^2 = \left(\frac{x}{a}\right)^2$, or $\frac{x^2}{a^2} - \frac{y^2}{b^2} = 1$,

which is a graph of hyperbola.

Ans. (D)

37. If $f(x) = x^2 - 2x + 4$, what must the value of k be equal to, so that the graph of $f(x + k)$ will be symmetric to the y-axis?

(A) 1

(B) $-\dfrac{1}{2}$

(C) 0

(D) $-\dfrac{1}{4}$

(E) $\dfrac{1}{4}$

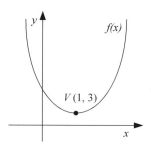

$f(x) = x^2 - 2x + 4 = (x - 1)^2 + 3$, thus, the vertex v = (1, 3) or using graphing calculator, we get the graph as shown. Since we want to get the graph to be symmetric to y-axis, we need to move our vertex (1, 3) to (0, 3) on y-axis. That is, moving the graph horizontally 1 unit to the left, or $f(x) \to f(x + 1)$, that is, $k = 1$.

Ans. (A)

38. If $\log_r 5 = A$ and $\log_r 15 = B$, then $\log_r \dfrac{r}{9}$ is equal to

(A) $1 - A + B$
(B) $1 + 2A - 2B$
(C) $1 + A - B$
(D) $1 - 2A + 2B$
(E) zero, if $r = 4$

To solve this problem, we need to refer to the properties of log function as in:

(1) $\log x \cdot y = \log x + \log y$

(2) $\log \dfrac{x}{y} = \log x - \log y$

(3) $\log x^m = m \cdot \log x$

(4) $\log_x y = \dfrac{\log y}{\log x} = \dfrac{\ln y}{\ln x}$

(5) $\log_x x = 1$, $\log_x 1 = 0$

(6) $y = \log_a x$, if and only if $x = a^y$

(7) $a^{\log_x y} = y^{\log_x a}$

Now, using the above properties,

$\log_r \dfrac{r}{9} = \log_r r - \log_r 9 = 1 - \log_r 3^2 = 1 - 2\log_r 3$.

But $\log_r 3 = \log_r \dfrac{15}{5} = \log_r 15 - \log_r 5 = B - A$.

Therefore, $\log_r \dfrac{r}{9} = 1 - 2(\log_r 3) = 1 - 2(B - A) = 1 + 2A - 2B$.

Ans. (B)

39. $\tan \left(\text{arc sin } \dfrac{3}{5}\right)$ equals

(A) 0.95
(B) 0.75
(C) 0.33
(D) 0.35
(E) 0.50

To get the value of $\tan \left(\sin^{-1}\dfrac{3}{5}\right)$, we need to get first,

$\sin^{-1}\dfrac{3}{5} = 36.9°$, by using calculator.

Then, $\tan \left(\sin^{-1}\dfrac{3}{5}\right) = \tan 36.9° = 0.75$

Ans. (B)

40. If a geometric sequence begins with the terms $\frac{1}{4}$, 2, ...,
what is the sum of the first 7 terms?

(A) $74898\frac{1}{4}$

(B) 56561

(C) $9362\frac{1}{4}$

(D) $1170\frac{1}{4}$

(E) 6

For a geometric sequence, we have formulas for a general term, $a_n = a_1 r^{n-1}$, and also for the sum, $s_n = a_1 \times \frac{1-r^n}{1-r}$, where r is the common ratio, and a_1 is the first term.
From the given terms, $\frac{1}{4}$, 2, ...,
we get $r = \frac{a_2}{a_1} = \frac{2}{\frac{1}{4}} = 8$, thus, $s_n = \frac{1}{4}\left(\frac{1-8^7}{1-8}\right) = 74898.25$.

Ans. (A)

41. If $\arccos(\sin x) = \frac{\pi}{6}$ and $0 \le x \le \frac{\pi}{2}$, then x could equal

(A) 0

(B) $\frac{\pi}{6}$

(C) $\frac{\pi}{4}$

(D) $\frac{\pi}{3}$

(E) $\frac{\pi}{2}$

$\cos^{-1}(\sin x) = \frac{\pi}{6}$, $0 \le x \le \frac{\pi}{2}$.
By applying cos on both sides of equation,
we get $\cos(\cos^{-1}(\sin x)) = \cos\frac{\pi}{6} = \frac{\sqrt{3}}{2}$.
Here, we may cancel cos and $\cos^{-1}$,
and get $\sin x = \frac{\sqrt{3}}{2}$, or $x = \sin^{-1}\frac{\sqrt{3}}{2} = 60°$ or $\frac{\pi}{3}$.

Ans. (D)

42. If vector $\vec{v} = (\sqrt{2}, \sqrt{3})$ and vector $\vec{u} = (1, -1)$,
find the value of $|\sqrt{3}\,\vec{v} + 3\,\vec{u}|$.

(A) 52
(B) $3 + \sqrt{6}$
(C) 6
(D) $3 - \sqrt{6}$
(E) 3

To solve this problem, we need to know the following properties of vector-operation:
Let vector $\vec{v} = (v_1, v_2)$ and $\vec{u} = (u_1, u_2)$, then

(1) $\vec{u} + \vec{v} = (u_1 \pm v_1, u_2 \pm v_2)$
(2) $c\vec{u} = (cu_1, cu_2)$
(3) magnitude of $\vec{u}$ or $|\vec{u}| = \sqrt{u_1^2 + u_2^2}$
(4) Unit vector $\vec{w}$ of $\vec{v}$ is, $\vec{w} = \frac{\vec{v}}{|\vec{v}|}$

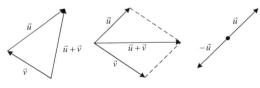

(5)

Now, given $\vec{v} = (\sqrt{2}, \sqrt{3})$, $\vec{u} = (1, -1)$
We get, $\sqrt{3}\,\vec{v} = (\sqrt{3}\sqrt{2}, \sqrt{3}\sqrt{3}) = (\sqrt{6}, 3)$ and $3\,\vec{u} = (3, -3)$
such that $\sqrt{3}\,\vec{v} + 3\,\vec{u} = (3 + \sqrt{6}, 0)$.
Therefore, using the property (3),
we get $|\sqrt{3}\,\vec{v} + 3\,\vec{u}| = \sqrt{(3 + \sqrt{6})^2 + 0^2} = 3 + \sqrt{6}$.

Ans. (B)

43. If $\tan A = \cot B$, which of the following must be true?

(A) $A = B$
(B) $A + B = 0$
(C) $A + B = 90°$
(D) $A + B = 180°$
(E) $A - B = 180°$

We need to remember the property of "Co-Function" in our Trig. Course, that is:

(1) $\sin \theta = \cos (90° - \theta) = \cos (\frac{\pi}{2} - \theta)$, (ie) $\sin 30° = \cos 60°$

(2) $\tan \theta = \cot (90° - \theta) = \cot (\frac{\pi}{2} - \theta)$, (ie) $\tan 15° = \cot 75°$

(3) $\sec \theta = \csc (90° - \theta) = \csc (\frac{\pi}{2} - \theta)$, (ie) $\sec 45° = \csc 45°$

Since $\tan A = \cot B$, by the property (2),
we must get $A + B = 90°$.

Ans. (C)

44. If the mean of the set of data 2, 3, 3, 7, 1, 5, x is $4.\overline{271}$, what is the value of x?

(A) 8.9
(B) -10.7
(C) 5.6
(D) 2.5
(E) 7.4

The mean can be calculated as,

$\dfrac{\text{sum}}{\text{\# of items}} = (2 + 3 + 3 + 7 + 1 + 5 + x) = 7 \times 4.\overline{271}$.

$\therefore x = 8.9$

Ans. (A)

45. A committee of 4 people is to be selected from 5 men and 7 women. If the selection is made randomly, what is the probability that the committee consists of 2 men and 2 women?

(A) $\dfrac{1}{3}$

(B) $\dfrac{14}{33}$

(C) $\dfrac{35}{144}$

(D) $\dfrac{4}{35}$

(E) $\dfrac{35}{495}$

There are total of 12 people,
consisted of 5 men and 7 women.
If we want to select 4 people out of 12 people, regardless of their sexes, then the total combination is $T = {}_{12}C_4$.
But the combination of selecting 2 men from 5 men is $T_m = {}_5C_2$, while the combination of selecting 2 women from 7 women is $T_w = {}_7C_2$.
Since selecting 2 men and 2 women happens simultaneously, we must multiply the two events,
and that is, $T_m \times T_w = {}_5C_2 \times {}_7C_2$.
Therefore, the probability that the committee consists of 2 men and 2 women will be,

$P = \dfrac{T_m \times T_w}{T} = \dfrac{{}_5C_2 \times {}_7C_2}{{}_{12}C_4} = \dfrac{14}{33}$.

Ans. (B)

46. The graph of $y = \log_2 \dfrac{1}{x}$ and $y = \ln \dfrac{x^2}{3}$ intersect at a point where x equals

(A) 6.24
(B) 1.38
(C) 1.69
(D) 1.05
(E) 5.44

$y_1 = \log_2 \dfrac{1}{x} = \log_2 x^{-1} = -\log_2 x$, or $y_1 = -\dfrac{\ln x}{\ln 2}$.

Also, $y_2 = \ln \dfrac{x^2}{3} = \ln x^2 - \ln 3 = 2\ln x - \ln 3$.

Now, using graphing calculator by setting, $y_1 = -\dfrac{\ln x}{\ln 2}$, and $y_2 = 2\ln x - \ln 3$, we get the intersection at $x = 1.38$.

Ans. (B)

47. Which of the following is equivalent to $\sin(\theta - \dfrac{\pi}{3}) + \cos(\theta - \dfrac{\pi}{6})$ for all values of θ?

(A) $\sin \theta$
(B) $\cos \theta$
(C) $\sqrt{3} \sin \theta + \cos \theta$
(D) $\sqrt{3} \sin \theta$
(E) $\sqrt{3} \cos \theta$

We may use a graphing calculator for this problem.
Let's graph $y_1 = \sin(\theta - \dfrac{\pi}{3}) + \cos(\theta - \dfrac{\pi}{6})$, and then put the answer choice (A) as y_2, or $y_2 = \sin \theta$.
You may find the graph of y_2 being exactly the same as the graph of y_1, which means, $y_1 = y_2$.

Ans. (A)

(Note) We may use the property of the angle sum and difference from the Trig function.

(1) $\sin(A \pm B) = \sin A \cos B \pm \cos A \sin B$
(2) $\cos(A \pm B) = \cos A \cos B \mp \sin A \sin B$
(3) $\tan(A \pm B) = \dfrac{\tan A \pm \tan B}{1 \mp \tan A \tan B}$
(4) $\sin 2A = 2 \sin A \cos A$
(5) $\cos 2A = \cos^2 A - \sin^2 A = 1 - 2\sin^2 A = 2\cos^2 A - 1$

But it is not necessary here, because it takes a much longer time.

48. What is the probability that a prime number is less than 11, given that it is less than 19?

(A) $\dfrac{1}{2}$ (B) $\dfrac{3}{4}$ (C) $\dfrac{4}{7}$ (D) $\dfrac{5}{8}$ (E) $\dfrac{2}{3}$

The definition of a prime number is the positive integer which has factors of 1 and itself only, such as 2, 3, 5, 7, ...
(please note that 1 is not a prime number)
Now, based on this definition, the prime number less than 11, becomes 2, 3, 5, and 7 only, while the prime numbers less than 19 are 2, 3, 5, 7, 11, 13 and 17.
Therefore, the probability becomes,
$p = \dfrac{4 \text{ numbers}}{7 \text{ numbers}} = \dfrac{4}{7}$.

Ans. (C)

49. If $(\csc x)(\cot x) < 0$, which of the following must be true?

I. $\tan x < 0$
II. $(\sin x)(\cos x) < 0$
III. x is in the second or third quadrant

(A) I only
(B) II only
(C) III only
(D) II and III
(E) I and II

Remember that $\csc x = \dfrac{1}{\sin x}$ and $\cot x = \dfrac{1}{\tan x}$.
Therefore, $(\csc x)(\cot x) = \dfrac{1}{\sin x} \times \dfrac{1}{\tan x} < 0$, which implies that the signs of $\sin x$ and $\tan x$ must be the opposite signs each other. Referring to the sign chart by quadrant,

(II)
sin (+)
cos (-)
tan (-)

(I)
All

(III)
tan (+)
cos (-)
sin (-)

(IV)
cos (+)
tan (-)
sin (-)

we have $\sin x$ and $\tan x$ have opposite signs each other in (II) and (III) quadrants. Therefore, the answer is (C).

Ans. (C)

50. If the graph below represents the function $f(x)$, which of the following could represent the equation of the inverse of f?

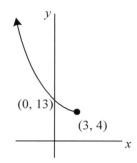

(0, 13)

(3, 4)

(A) $x = y^2 - 7y - 2$
(B) $x = y^2 + 2$
(C) $x = (y - 3)^2 + 4$
(D) $x = (y + 2)^2 - 3$
(E) $x = (y + 4)^2 + 3$

For this problem, we need to understand the following properties of inverse function $f^{-1}(x)$:

1. The function has to be one to one function.
2. $f(x)$ and $f^{-1}(x)$ are symmetric to each other on $y = x$ line.
3. To get $f^{-1}(x)$, we need to switch x into y, and y into x.
4. The domain of $f(x)$ becomes the range of $f^{-1}(x)$, and the range of $f(x)$ becomes the domain of $f^{-1}(x)$.

Based on these properties, we need to figure out first what the original function, $f(x)$ is, as given in the problem. $f(x)$ has vertex point at $(3, 4)$ and it has y-intercept of $(0, 13)$. Refer to the standard form with vertex at (h, k), $y = a(x - h)^2 + k$.
Now, plug in $h = 3$, $k = 4$, and also the consider the coordinates of the y-intercept at $x = 0$, $y = 13$.
Then, we get $13 = a(0 - 3)^2 + 4$, or $a = 1$.
Therefore, our original $f(x)$ becomes,
$f(x) = y = (x - 3)^2 + 4$, $x \leq 3$, $y \geq 4$.
Since the inverse function $f^{-1}(x)$ is by switching x and y's, we get $f^{-1}(x)$, or $x = (y - 3)^2 + 4$, $y \leq 3$, $x \geq 4$.

Ans. (C)

Model Test No. 02

50 Questions / 60 Minutes

Directions: For each question, determine which of the answer choices is correct and fill in the oval on the answer sheet that corresponds to your choice.

Notes:

1. You will need to use a scientific or graphing calculator to answer some of the questions.
2. Be sure your calculator is in degree mode.
3. Each figure on this test is drawn as accurately as possible unless it is specifically indicated that the figure has not been drawn to scale.
4. The domain of any function f is the set of all real numbers x for which $f(x)$ is also a real number, unless the question indicates that the domain has been restricted in some way.
5. The box below contains five formulas that you may need to answer one or more of the questions.

REFERENCE INFORMATION

THE FOLLOWING INFORMATION IS FOR YOUR REFERENCE IN ANSWERING SOME OF THE QUESTIONS IN THIS TEST.

Volume of a right circular cone with radius r and height h: $V = \frac{1}{3}\pi r^2 h$

Lateral Area of a right circular cone with circumference of the base c and slant height l: $S = \frac{1}{2}cl$

Volume of a sphere with radius r: $V = \frac{4}{3}\pi r^3$

Surface Area of a sphere with radius r: $S = 4\pi r^2$

Volume of a pyramid with base area B and height h: $V = \frac{1}{3}Bh$

Answer Sheet
Model Test No. 02

1 Ⓐ Ⓑ Ⓒ Ⓓ Ⓔ	14 Ⓐ Ⓑ Ⓒ Ⓓ Ⓔ	27 Ⓐ Ⓑ Ⓒ Ⓓ Ⓔ	40 Ⓐ Ⓑ Ⓒ Ⓓ Ⓔ
2 Ⓐ Ⓑ Ⓒ Ⓓ Ⓔ	15 Ⓐ Ⓑ Ⓒ Ⓓ Ⓔ	28 Ⓐ Ⓑ Ⓒ Ⓓ Ⓔ	41 Ⓐ Ⓑ Ⓒ Ⓓ Ⓔ
3 Ⓐ Ⓑ Ⓒ Ⓓ Ⓔ	16 Ⓐ Ⓑ Ⓒ Ⓓ Ⓔ	29 Ⓐ Ⓑ Ⓒ Ⓓ Ⓔ	42 Ⓐ Ⓑ Ⓒ Ⓓ Ⓔ
4 Ⓐ Ⓑ Ⓒ Ⓓ Ⓔ	17 Ⓐ Ⓑ Ⓒ Ⓓ Ⓔ	30 Ⓐ Ⓑ Ⓒ Ⓓ Ⓔ	43 Ⓐ Ⓑ Ⓒ Ⓓ Ⓔ
5 Ⓐ Ⓑ Ⓒ Ⓓ Ⓔ	18 Ⓐ Ⓑ Ⓒ Ⓓ Ⓔ	31 Ⓐ Ⓑ Ⓒ Ⓓ Ⓔ	44 Ⓐ Ⓑ Ⓒ Ⓓ Ⓔ
6 Ⓐ Ⓑ Ⓒ Ⓓ Ⓔ	19 Ⓐ Ⓑ Ⓒ Ⓓ Ⓔ	32 Ⓐ Ⓑ Ⓒ Ⓓ Ⓔ	45 Ⓐ Ⓑ Ⓒ Ⓓ Ⓔ
7 Ⓐ Ⓑ Ⓒ Ⓓ Ⓔ	20 Ⓐ Ⓑ Ⓒ Ⓓ Ⓔ	33 Ⓐ Ⓑ Ⓒ Ⓓ Ⓔ	46 Ⓐ Ⓑ Ⓒ Ⓓ Ⓔ
8 Ⓐ Ⓑ Ⓒ Ⓓ Ⓔ	21 Ⓐ Ⓑ Ⓒ Ⓓ Ⓔ	34 Ⓐ Ⓑ Ⓒ Ⓓ Ⓔ	47 Ⓐ Ⓑ Ⓒ Ⓓ Ⓔ
9 Ⓐ Ⓑ Ⓒ Ⓓ Ⓔ	22 Ⓐ Ⓑ Ⓒ Ⓓ Ⓔ	35 Ⓐ Ⓑ Ⓒ Ⓓ Ⓔ	48 Ⓐ Ⓑ Ⓒ Ⓓ Ⓔ
10 Ⓐ Ⓑ Ⓒ Ⓓ Ⓔ	23 Ⓐ Ⓑ Ⓒ Ⓓ Ⓔ	36 Ⓐ Ⓑ Ⓒ Ⓓ Ⓔ	49 Ⓐ Ⓑ Ⓒ Ⓓ Ⓔ
11 Ⓐ Ⓑ Ⓒ Ⓓ Ⓔ	24 Ⓐ Ⓑ Ⓒ Ⓓ Ⓔ	37 Ⓐ Ⓑ Ⓒ Ⓓ Ⓔ	50 Ⓐ Ⓑ Ⓒ Ⓓ Ⓔ
12 Ⓐ Ⓑ Ⓒ Ⓓ Ⓔ	25 Ⓐ Ⓑ Ⓒ Ⓓ Ⓔ	38 Ⓐ Ⓑ Ⓒ Ⓓ Ⓔ	
13 Ⓐ Ⓑ Ⓒ Ⓓ Ⓔ	26 Ⓐ Ⓑ Ⓒ Ⓓ Ⓔ	39 Ⓐ Ⓑ Ⓒ Ⓓ Ⓔ	

1. The positive zero of $y = x^2 + 2x - \dfrac{3}{5}$ is, to the nearest tenth, equal to

   (A) 0.8
   (B) $0.7 + 1.1i$
   (C) 0.7
   (D) 0.3
   (E) 2.2

2. The solution set of $x^2 - 2x < 8$ is given by the inequality

   (A) $x < 4$
   (B) $x > -2$
   (C) $-2 < x < 4$
   (D) $-2 \le x \le 4$
   (D) $x > 4$

3. If $\sin x = 0.6018$, then $\sec x =$

   (A) 2.1290
   (B) 1.2521
   (C) 1.0818
   (D) 0.9243
   (E) 0.4890

4. An equation of line l in Figure 6 is

   (A) $x = 2$
   (B) $y = 2$
   (C) $x = 0$
   (D) $y = x + 2$
   (E) $x + y = 2$

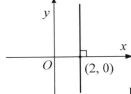

   Figure 6

5. If $0 < x < 1$, then

   (A) $0 < \ln \dfrac{1}{x} < 1$

   (B) $\ln \dfrac{1}{x} > 1$

   (C) $\ln \dfrac{1}{x} < 0$

   (D) $\ln \dfrac{1}{x} > 0$

   (E) none of these is true

6. When $(a + b)^4$ is expanded,
   what is the coefficient of the third term?

   (A) 10
   (B) 11
   (C) 6
   (D) -11
   (E) The sum cannot be determined.

7. What is the approximate x-intercept of $f(x) = \dfrac{x+\sqrt{3}}{x-3}$?

   (A) -1.73
   (B) 1.73
   (C) -0.58
   (D) 0.58
   (E) 3

8. If $f(x) = \sqrt{x + 4}$ and $g(x) = 2x^2 - 3$, then $f(g(2)) =$

   (A) 6.16
   (B) 3.61
   (C) 2.24
   (D) 3.00
   (E) 6.00

9. Let the symbol $\triangle$ be defined as $A \triangle B = \cos A \cdot \sin B - \sin A \cdot \cos B$. What is the approximate value of $76° \triangle 36°$?

   (A) .766
   (B) .643
   (C) .125
   (D) 0
   (E) -.643

10. The slope of a line perpendicular to the line
    whose equation is $\dfrac{x}{4} + \dfrac{y}{3} = 1$ is

    (A) $\dfrac{1}{4}$

    (B) $-\dfrac{4}{3}$

    (C) $-\dfrac{3}{4}$

    (D) $\dfrac{4}{3}$

    (E) -3

70

11. Approximately, what is $\lim\limits_{x\to\infty} \dfrac{3\sqrt{5}x^3-5}{2x^3+3x^2+4}$?

(A) 3.02
(B) 3.13
(C) 3.24
(D) 3.35
(E) 3.46

12. The maximum value of $3\cdot\sin x\cdot\cos x + \dfrac{3}{2}$ is

(A) $\dfrac{3}{2}$

(B) $\dfrac{5}{2}$

(C) 3

(D) $\dfrac{7}{2}$

(E) 6

13. As angle x increases from 0 to 2π radians, $\sin x$ increases in

(A) no quadrants
(B) the first and third quadrants only
(C) the second and fourth quadrants only
(D) all four quadrants
(E) the first and fourth quadrants only

14. The solution set of $2x - 5y < 3$ lies in which quadrants?

(A) I only
(B) I and II
(C) I, II, and III
(D) II, III, and IV
(E) I, II, III, and IV

15. If $f(x) = 2\cdot e^x - 4$ and $g(x) = \ln x$, then $f(g(9)) =$

(A) 6.83
(B) 12
(C) 14
(D) 45.98
(E) 568.17

16. Find the positive value of cos (sin$^{-1}\frac{1}{2}$).

    (A) $\frac{1}{2}$

    (B) 1

    (C) $\frac{\sqrt{3}}{2}$

    (D) $\sqrt{3}$

    (E) 2

17. How many different hands, each consisting of four cards, can be drawn, from a deck of 52 different cards?

    (A) 132,600
    (B) 1.3×10^{67}
    (C) 270,725
    (D) 1.4×10^{10}
    (E) 2652

18. The area of a triangle with sides 4, 5, and 8 is

    (A) 7.5
    (B) 8.18
    (C) 3.75
    (D) 13.0
    (E) 2.4

19. One side of a given triangle is 6 inches. Inside the triangle a line segment is drawn parallel to this side, cutting off a triangle whose area is one-thirds that of the given triangle. Find the length of this segment in inches.

    (A) 2
    (B) $2\sqrt{3}$
    (C) $3\sqrt{3}$
    (D) $6\sqrt{3}$
    (E) 9

20. Given the statement, "Only if it snows will the class be cancelled", if it is known that it did not snow, what conclusion can be drawn?

    (A) The class is cancelled.
    (B) The class is not cancelled.
    (C) The class might be cancelled.
    (D) No conclusion can be drawn.
    (E) None of the above.

USE THIS SPACE FOR SCRATCH WORK

72

21. What is the domain of the function $f(x) = \dfrac{\sqrt{2x^3+5}}{\sqrt{2x^3-5}}$?

    (A) $x < -1.36$
    (B) $-1.36 \leq x \leq 1.36$
    (C) $x > 1.36$
    (D) $x \geq 1.36$
    (E) $x < -1.36$ or $x > 1.36$

22. The hyperbola $9x^2 - 4y^2 = 16$ intersects the x-axis at approximately which of the following points?

    (A) $(2, 0)$
    (B) $(0, 2)$
    (C) $(1.33, 0)$
    (D) $(0, 1.33)$
    (E) $(0, -1.33)$

23. The length of the vector that could correctly be used to represent in the complex plane the number $-\sqrt{7} + 2i$ is

    (A) 11
    (B) $\sqrt{11}$
    (C) $3\sqrt{2}$
    (D) $\sqrt{5}$
    (E) $\sqrt{13}$

24. In a cube, the ratio of the longest diagonal to a side is

    (A) $\sqrt{2} : \sqrt{3}$
    (B) $\sqrt{6} : 2$
    (C) $\sqrt{6} : \sqrt{2}$
    (D) $\sqrt{3} : \sqrt{2}$
    (E) $\sqrt{3} : 1$

25. What is the measure of the smallest angle in a right triangle with sides of lengths 5, 12 and 13?

    (A) $21.04°$
    (B) $22.62°$
    (C) $24.62°$
    (D) $42.71°$
    (E) $67.34°$

26. What is the period of the curve whose equation is $y = 4\sin x \cdot \cos x$?

    (A) $60°$
    (B) $120°$
    (C) $180°$
    (D) $360°$
    (E) $720°$

27. Let $f(x)$ be a polynomial function.
    If $f(\sqrt{2}) = 0$ and $f(-\sqrt{2}) = 0$, then $f(x)$ is divisible by

    (A) $x - 3$
    (B) $x^2 - 2$
    (C) $x^2 + 2$
    (D) $x^2 - 3x + 2$
    (E) $x^2 + 3x + 2$

28. How many real roots does the following equation have?
    $5e^x + e^{-x} - 4 = 0$

    (A) 0
    (B) 1
    (C) 2
    (D) 4
    (E) an infinite number

29. If $f(x) = \ln x$ and $g(x) = f(x) \cdot f^{-1}(x)$,
    what does $g(2)$ equal?

    (A) 8.1
    (B) 7.5
    (C) 8.3
    (D) 5.1
    (E) 7.4

30. If the 3rd term of a geometric progression is $\sqrt[3]{a}$
    and the 6th term is $\sqrt[3]{a^2}$,
    what is the 12th term of the progression?

    (A) a

    (B) $2a$

    (C) $a^{\frac{4}{3}}$

    (D) $a^{\frac{5}{3}}$

    (E) a^2

31. If $f(x, y, z) = \sqrt{3}x - \sqrt{5}y + \sqrt{6}z$ and
    $f(a, b, 0) = f(0, a, b)$, then $\dfrac{a}{b} =$

    (A) 0.32
    (B) 2.7
    (C) 8.6
    (D) 0.12
    (E) 1.18

32. What is the inverse of $f(x) = \dfrac{1}{\sqrt{-x^2+5}}$?

 (A) $y = \dfrac{1}{\sqrt{5-x^2}}$

 (B) $y = 5 - \dfrac{1}{x^2}$

 (C) $y = \sqrt{5 - \dfrac{1}{x^2}}$

 (D) $y = \sqrt{5 + \dfrac{1}{x^2}}$

 (E) $y = \sqrt{-x^2 + 5}$

33. Find the value of P if $(x + 1)$ is a factor of
 $f(x) = 3x^4 + 2x^3 - Px^2 - 2x + 1$.

 (A) 1
 (B) 2
 (C) 3
 (D) 4
 (E) 5

34. In how many different orders can 10 students arrange themselves around a table?

 (A) 9
 (B) 81
 (C) 181,440
 (D) 362,880
 (E) 387,420,489

35. If x and y are real numbers,
 which one of the following relations is a function of x?

 (A) $\{(x, y) \mid y = \mid x \mid\}$
 (B) $\{(x, y) \mid x = y^2 - 2y + 4\}$
 (C) $\{(x, y) \mid y = \pm\sqrt{1 - x}\}$
 (D) $\{(x, y) \mid y < x - 2\}$
 (E) $\{(x, y) \mid x = \sec y\}$

36. If the equation $3x^3 + 9x^2 + px - q = 0$ has 3 equal roots, then

 (A) $q = 0$
 (B) $p = 0$
 (C) $q = 1$
 (D) each root $= 1$
 (E) each root $= -1$

75

37. Each term of a sequence, after the first, is inversely proportional to the term preceding it. If the first two terms are 3 and 9, what is the 100th term?

(A) 3
(B) 9
(C) 3^{100}
(D) 3^{99}
(E) The 100th term cannot be determined.

38. The point whose polar coordinates are $(3, 45°)$ is the same as the point whose polar coordinates are

(A) $(-3, 45°)$
(B) $(-3, 225°)$
(C) $(3, -45°)$
(D) $(-3, 135°)$
(E) $(3, 225°)$

39. If $\log x^2 \geq \log 8 + \dfrac{1}{2} \log x$, then

(A) $x \geq 2$
(B) $x \leq 2$
(C) $x \leq 4$
(D) $x \geq 4$
(E) $x \geq 1$

40. A right circular cylinder is circumscribed about a sphere. If V_s represents the volume of the sphere and V_c represents the volume of the cylinder, then

(A) $V_s = \dfrac{2}{3}V_c$

(B) $V_s < \dfrac{2}{3}V_c$

(C) $V_s > \dfrac{2}{3}V_c$

(D) $V_s \leq \dfrac{2}{3}V_c$

(E) $V_s \geq \dfrac{2}{3}V_c$

76

41. $f(x) = \begin{cases} \dfrac{x^2 - 2x - 3}{x - 3}, & \text{when } x \neq 3 \\ \\ k, & \text{when } x = 3, \end{cases}$

what must the value of k be in order for $f(x)$ to be a continuous function?

(A) 2
(B) 4
(C) 0
(D) -2
(E) No value of k will make $f(x)$ a continuous function.

42. Write the complex number $\sqrt{3} - i$ in polar form.

(A) $2(\cos \dfrac{\pi}{6} + i \cdot \sin \dfrac{\pi}{6})$

(B) $-2(\cos \dfrac{\pi}{6} + i \cdot \sin \dfrac{\pi}{6})$

(C) $2(\cos \dfrac{7\pi}{6} + i \cdot \sin \dfrac{7\pi}{6})$

(D) $-2(\cos \dfrac{11\pi}{6} + i \cdot \sin \dfrac{11\pi}{6})$

(E) $2(\cos \dfrac{11\pi}{6} + i \cdot \sin \dfrac{11\pi}{6})$

43. If $\dfrac{\cos^2 \theta - \sin^2 \theta}{\sin \theta \cos \theta} = 2\sqrt{3}$, then $\theta =$

(A) 15°
(B) 30°
(C) 45°
(D) 60°
(E) 75°

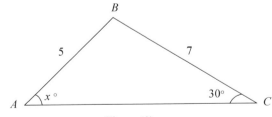

Figure (1)

44. What is the degree measure of the angle x of a triangle that has sides of length of 5 and 7 as shown in the figure (1)?

(A) 44.43°
(B) 54.31°
(C) 71.37°
(D) 25.69°
(E) 34.31°

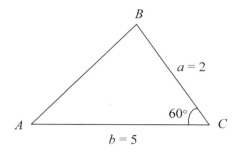

45. In $\triangle ABC$, $a = 2$, $b = 5$, $\angle C = 60°$.
What is the area of $\triangle ABC$?

(A) 4.3
(B) 3.6
(C) 3.2
(D) 2.9
(E) 2.3

46. If $f(x, y) = x^3 - xy^2 + 3y^4$,
which of the following is (are) true?

I. $f(x, y) = f(x, -y)$
II. $f(x, y) = f(-x, y)$
III. $f(x, y) = f(-x, -y)$

(A) I only
(B) II only
(C) III only
(D) I and III only
(E) I, II, and III

47. Figure 5 shows a cube with edge of length 4 centimeters.
If points A and C are midpoints of the edges of the cube,
what is the area of region $ABCD$?

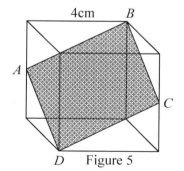

Figure 5

(A) 15.71 cm²
(B) 17.25 cm²
(C) 19.60 cm²
(D) 20.00 cm²
(E) 22.63 cm²

48. If (x, y) represents a point on the graph of $y = \sqrt{2x + 1}$, which of the following could be a portion of the graph of the set of points (x, y^2)?

(A)
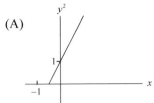

(B)
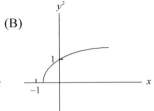

(C)
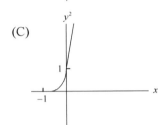

(D)
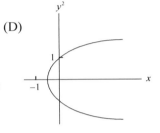

(E)
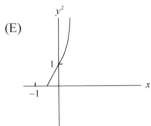

49. If the parameter is eliminated from the equations
$x = 4t^2 + 1$ and $y = 2t$,
then the relation between x and y is

(A) $y = x - 1$
(B) $y = 1 - x$
(C) $y^2 = x - 1$
(D) $y^2 = (x - 1)^2$
(E) $y^2 = 4x - 4$

50. If 8 people shook hands with each other, how many handshakes were exchanged?

(A) 8
(B) 16
(C) 21
(D) 28
(E) 56

Answer Key
Model Test No. 02

1	D	14	E	27	B	40	A
2	C	15	C	28	A	41	B
3	B	16	C	29	D	42	E
4	A	17	C	30	C	43	A
5	D	18	B	31	E	44	A
6	C	19	B	32	C	45	A
7	A	20	B	33	D	46	A
8	D	21	C	34	D	47	D
9	E	22	C	35	A	48	A
10	D	23	B	36	E	49	C
11	D	24	E	37	B	50	D
12	C	25	B	38	B		
13	E	26	C	39	D		

How to Score the SAT Subject Test in Mathematics Level 2

When you take an actual SAT Subject Test in Mathematics Level 2, your answer sheet will be "read" by a scanning machine that will record your responses to each question. Then a computer will compare your answers with the correct answers and produce your raw score. You get one point for each correct answer. For each wrong answer, you lose one-fourth of a point. Questions you omit (and any for which you mark more than one answer) are not counted. This raw score is converted to a scaled score that is reported to you and to the colleges you specify.

Finding Your Raw Test Score

STEP 1: Table A lists the correct answers for all the questions on the Subject Test in Mathematics Level 2 that is reproduced in this book. It also serves as a worksheet for you to calculate your raw score.

- Compare your answers with those given in the table.
- Put a check in the column marked "Right" if your answer is correct.
- Put a check in the column marked "Wrong" if your answer is incorrect.
- Leave both columns blank if you omitted the question.

STEP 2: Count the number of right answers.
Enter the total here: _____

STEP 3: Count the number of wrong answers.
Enter the total here: _____

STEP 4: Multiply the number of wrong answers by .250.
Enter the product here: _____

STEP 5: Subtract the result obtained in Step 4 from the total you obtained in Step 2.
Enter the result here: _____

STEP 6: Round the number obtained in Step 5 to the nearest whole number.
Enter the result here: _____

The number you obtained in Step 6 is your raw score.

Scaled Score Conversion Table
Subject Test in Mathematics Level 2

Raw Score	Scaled Score	Raw Score	Scaled Score	Raw Score	Scaled Score
50	800	28	630	6	470
49	800	27	630	5	460
48	800	26	620	4	450
47	800	25	610	3	440
46	800	24	600	2	430
45	800	23	600	1	420
44	800	22	590	0	410
43	790	21	580	-1	400
42	780	20	580	-2	390
41	770	19	570	-3	370
40	760	18	560	-4	360
39	750	17	560	-5	350
38	740	16	550	-6	340
37	730	15	540	-7	340
36	710	14	530	-8	330
35	700	13	530	-9	330
34	690	12	520	-10	320
33	680	11	510	-11	310
32	670	10	500	-12	300
31	660	9	490		
30	650	8	480		
29	640	7	480		

1. The positive zero of $y = x^2 + 2x - \dfrac{3}{5}$ is, to the nearest tenth, equal to

   (A) 0.8
   (B) $0.7 + 1.1i$
   (C) 0.7
   (D) 0.3
   (E) 2.2

Using Quadratic formula, $x = \dfrac{-b \pm \sqrt{b^2 - 4ac}}{2a}$, with $a = 1$, $b = 2$, $c = -\dfrac{3}{5}$, we get $x = 0.3$ for the positive root.

Ans. (D)

2. The solution set of $x^2 - 2x < 8$ is given by the inequality

   (A) $x < 4$
   (B) $x > -2$
   (C) $-2 < x < 4$
   (D) $-2 \le x \le 4$
   (D) $x > 4$

$x^2 - 2x - 8 < 0$, or $(x - 4)(x + 2) < 0$,
Therefore $-2 < x < 4$

Ans. (C)

3. If $\sin x = 0.6018$, then $\sec x =$

   (A) 2.1290
   (B) 1.2521
   (C) 1.0818
   (D) 0.9243
   (E) 0.4890

$\sin x = 0.618$, therefore, $x = \sin^{-1}(0.6018) = 37°$.
Now, $\sec x = \dfrac{1}{\cos x} = \dfrac{1}{\cos 37°} = 1.25$

Ans. (B)

4. An equation of line l in Figure 6 is

   (A) $x = 2$
   (B) $y = 2$
   (C) $x = 0$
   (D) $y = x + 2$
   (E) $x + y = 2$

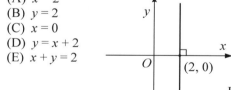

Figure 6

As shown in the figure, $x = 2$

Ans. (A)

5. If $0 < x < 1$, then

   (A) $0 < \ln \dfrac{1}{x} < 1$

   (B) $\ln \dfrac{1}{x} > 1$

   (C) $\ln \dfrac{1}{x} < 0$

   (D) $\ln \dfrac{1}{x} > 0$

   (E) none of these is true

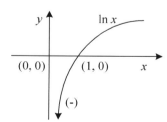

$0 < x < 1$, therefore, $\ln x < 0$.
But, $\ln \dfrac{1}{x} = \ln (x)^{-1} = (-1) \cdot \ln x$.
Since $\ln x < 0$, we get $\ln \dfrac{1}{x} = (-)(-) > 0$.

Ans. (D)

6. When $(a + b)^4$ is expanded, what is the coefficient of the third term?

    (A) 10
    (B) 11
    (C) 6
    (D) -11
    (E) The sum cannot be determined.

Refer to Binomial Expansion

$$\boxed{(a + b)^n = \sum_{r=0}^{n} {}_nC_r a^{n-r} b^r}.$$

Using $n = 4$, $r = 2$,
we get the coefficient of the third term, ${}_4C_2 = 6$.

Ans. (C)

7. What is the approximate x-intercept of $f(x) = \dfrac{x+\sqrt{3}}{x-3}$?

    (A) -1.73
    (B) 1.73
    (C) -0.58
    (D) 0.58
    (E) 3

To get the x – intercept,
we get $y = f(x) = 0$, $f(x) = \dfrac{x+\sqrt{3}}{x-3} = 0$.
Thus, $x + \sqrt{3} = 0$.
$\therefore x = -\sqrt{3} = $ -1.73.

Ans. (A)

8. If $f(x) = \sqrt{x + 4}$ and $g(x) = 2x^2 - 3$, then $f(g(2)) =$

    (A) 6.16
    (B) 3.61
    (C) 2.24
    (D) 3.00
    (E) 6.00

This problem is about composite function.
$f(g(2))$ means, first we get $g(2) = 2(2)^2 - 3 = 5$,
and then plug it into $f(x)$ function.
$\therefore f(5) = \sqrt{5 + 4} = \sqrt{9} = 3$

Ans. (D)

9. Let the symbol $\triangle$ be defined as $A \triangle B = \cos A \cdot \sin B - \sin A \cdot \cos B$. What is the approximate value of $76° \triangle 36°$?

    (A) .766
    (B) .643
    (C) .125
    (D) 0
    (E) -.643

Just replacing $A = 76°$, and $B = 36°$,
we get $\cos(76°) \cdot \sin(36°) - \sin(76°) \cdot \cos(36°) = (-)0.643$.

Ans. (E)

10. The slope of a line perpendicular to the line whose equation is $\dfrac{x}{4} + \dfrac{y}{3} = 1$ is

    (A) $\dfrac{1}{4}$

    (B) $-\dfrac{4}{3}$

    (C) $-\dfrac{3}{4}$

    (D) $\dfrac{4}{3}$

    (E) -3

$\dfrac{x}{4} + \dfrac{y}{3} = 1$ into a standard eq., we get $y = -\dfrac{3}{4}x + 3$.
Now, to be perpendicular to this line, we need the slope m,
which is a negative reciprocal of $-\dfrac{3}{4}$.

$$\therefore m = \dfrac{1}{-\left(-\dfrac{3}{4}\right)} = \dfrac{4}{3}.$$

Ans. (D)

11. Approximately, what is $\lim\limits_{x \to \infty} \frac{3\sqrt{5}x^3 - 5}{2x^3 + 3x^2 + 4}$?

    (A) 3.02
    (B) 3.13
    (C) 3.24
    (D) 3.35
    (E) 3.46

To find the limit of this eq., notice that the degrees of numerator and denominator are both the same degrees of 3. Thus, the limit on this will be the ratio of the highest degree coefficient, which is $\frac{3\sqrt{5}}{2} = 3.35$.

Ans. (D)

Limit for Horizontal Asymptotes & Vertical Asymptotes;

- Vertical Asymptotes is the vertical line that the curve is approaching, but never cross over as $x \to a$.
  In most cases, the rational function has its VA, when we find its denominator equals to zero.
- Horizontal Asymptotes is the horizontal line that the curve is approaching as $x \to \infty$.
  To find HA, we need to know 3 different cases.
  That is, HA, $y = \lim\limits_{x \to \infty} f(x)$.

  1. When the degree of Numerator = the degree of Denominator, then HA is just the ratio of the coef. of the highest degree terms.
  2. When the degree of the Num > the degree of denominator, then it is ZERO.
  3. When the degree of Num < the degree of Denom, then it is UNDEFINED.

- For example, let $y = \frac{x^2}{x^2 - 1}$, where $x \neq$ -1, 1.
  Then, since the degrees of numerator and denominator are both two, the HA is the limit value, which is the ratio of the leading coefficients.
  That is, our horizontal asymptote is $y = \frac{1}{1} = 1$.

Now, to get the vertical asymptotes, where the function is undefined, we set its denominator equal zero. That is, $x^2 - 1 = 0$. Our vertical asymptotes are $x = 1$, $x = $ -1.

(Note: Be careful about "removable discontinuity" !!) In general, for Math IIC test when time management is very important, we may just use a graphing calculator, which is a much more effective way to solve quickly during the exam. You may find from the graphing of $y = \frac{x^2}{x^2 - 1}$, two vertical asymptotes at $x = 1$, $x = $ -1, and one horizontal asymptote at $y = 1$.

12. The maximum value of $3 \cdot \sin x \cdot \cos x + \frac{3}{2}$ is

(A) $\frac{3}{2}$

(B) $\frac{5}{2}$

(C) 3

(D) $\frac{7}{2}$

(E) 6

Using graphing calculator, we get max = 3.

Or, since $3\sin x \cdot \cos x + \frac{3}{2} = \frac{3}{2}(2\sin x \cdot \cos x) + \frac{3}{2}$,

and $2\sin x \cdot \cos x = \sin 2x$, we get $\frac{3}{2}(\sin 2x) + \frac{3}{2}$.

The amplitude of this function is $\frac{3}{2}$,

but the function has been shifted up by $\frac{3}{2}$.

Therefore, the maximum value becomes $(\frac{3}{2} + \frac{3}{2}) = 3$.

Ans. (C)

Trigonometry

A) 6 Basic Trig Functions;

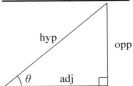

- $\sin \theta = \dfrac{\text{opp}}{\text{hyp}}$, $\cos \theta = \dfrac{\text{adj}}{\text{hyp}}$, $\tan \theta = \dfrac{\text{opp}}{\text{adj}}$ ...
  SOH-CAH-TOA!!!
- $\csc \theta = \dfrac{1}{\sin \theta}$, $\sec \theta = \dfrac{1}{\cos \theta}$, $\cot \theta = \dfrac{1}{\tan \theta}$

B) The property of the angle sum and difference

- $\sin (A \pm B) = \sin A \cos B \pm \cos A \sin B$
- $\cos (A \pm B) = \cos A \cos B \mp \sin A \sin B$
- $\tan (A \pm B) = \dfrac{\tan A \pm \tan B}{1 \mp \tan A \tan B}$
- $\sin 2A = 2 \sin A \cos A$
- $\cos 2A = \cos^2 A - \sin^2 A = 1 - 2\sin^2 A = 2\cos^2 A - 1$
- $\sin^2 A + \cos^2 A = 1$
- $1 + \tan^2 \theta = \sec^2 \theta$.

C) Heron's formula to find the area, given 3 sides;

Area $= \sqrt{s(s-a)(s-b)(s-c)}$, where $s = \dfrac{a+b+c}{2}$

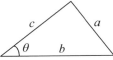

- The Law of Sine: $\dfrac{\sin A}{a} = \dfrac{\sin B}{b} = \dfrac{\sin C}{c}$
- The law of Cosine: $c^2 = a^2 + b^2 - 2ab \cdot \cos C$
- Area with 2 side, and angle θ: Area $= \dfrac{1}{2} bc \cdot \sin \theta$

D) For a general form: $y = a \cos b(x - h) + k$, we have,

$|a|$ = amplitude, b = # of frequency, $\dfrac{2\pi}{b}$ = period,

h = units of translation horizontally,

k = units of shifting vertically.

F) <u>Sign changes by Quadrants:</u>
all seniors take calculus…
by I-II-III-IV Qtr

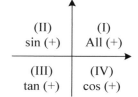

13. As angle x increases from 0 to 2π radians, $\sin x$ increases in

    (A) no quadrants
    (B) the first and third quadrants only
    (C) the second and fourth quadrants only
    (D) all four quadrants
    (E) the first and fourth quadrants only

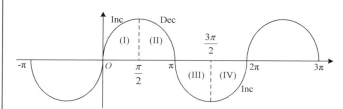

As shown in this graph, $y = \sin x$ increases in (I) and (IV).

Ans. (E)

14. The solution set of $2x - 5y < 3$ lies in which quadrants?

    (A) I only
    (B) I and II
    (C) I, II, and III
    (D) II, III, and IV
    (E) I, II, III, and IV

$2x - 5y < 3$, or $2x - 3 < 5y \rightarrow \dfrac{2}{5}x - \dfrac{3}{5} < y$
As shown here, the graph covers all four quadrants.

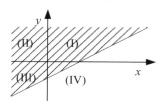

Ans. (E)

15. If $f(x) = 2 \cdot e^x - 4$ and $g(x) = \ln x$, then $f(g(9)) =$

    (A) 6.83
    (B) 12
    (C) 14
    (D) 45.98
    (E) 568.17

As in question #8, we get $g(9) = \ln 9$, and then
$f(g(9)) = f(\ln 9) = 2 \cdot (9^{\ln e}) - 4 = 18 - 4 = 14$. (note: $a^{\ln b} = b^{\ln a}$)

Ans. (C)

<u>Properties of log function</u>:
(1) $\log x \cdot y = \log x + \log y$
(2) $\log \dfrac{x}{y} = \log x - \log y$
(3) $\log x^m = m \cdot \log x$
(4) $\log_x y = \dfrac{\log y}{\log x} = \dfrac{\ln y}{\ln x}$
(5) $\log_x x = 1$, $\log_x 1 = 0$
(6) $y = \log_a x$, if and only if $x = a^y$

16. Find the positive value of cos $(\sin^{-1} \frac{1}{2})$.

(A) $\frac{1}{2}$ (B) 1   (C) $\frac{\sqrt{3}}{2}$ (D) $\sqrt{3}$ (E) 2

Here, $\sin^{-1} \frac{1}{2} = 30°$. $\therefore \cos(30°) = \frac{\sqrt{3}}{2}$.

Ans. (C)

17. How many different hands, each consisting of four cards,
can be drawn, from a deck of 52 different cards?

(A) 132,600
(B) 1.3×10^{67}
(C) 270,725
(D) 1.4×10^{10}
(E) 2652

To solve this problem, we need to understand the following ideas on probability and combination.

1) The total possible number of combination on n objects is $n!$
(eg) (A, B, C) = 3!, because A-B-C, A-C-B, B-A-C, B-C-A, C-A-B, C-B-A, total of 6 = 3!
(A, B, C, D) = 4!, (A, B, C, D, E) = 5! … etc.

2) The total number of combination on n objects, but with the repetition of k and r objects is $\frac{n!}{k!r!}$

(eg) COFFEE $\rightarrow \frac{6 \text{ letters}}{(2F's)(2E's)} = \frac{6!}{2!2!}$

Minimum $\rightarrow \frac{7 \text{ letters}}{(3m's)(2i's)} = \frac{7!}{3!2!}$

3) The total number of combination on n objects, but with k group with n, and n_2… member in them is $k!(n_1!)(n_2!)…$
(eg) A, B, C, D, E into 2 grouping of (A, B), (C, D, E) will be (2! grouping)(2! on A, B)(3! on C, D, E) = (2!)(2!)(3!)

4) Permutation is with a consideration of "Ordering",
$_nP_r = \frac{n!}{(n-r)!}$
(eg) Suggest I choose 3 students with the best score out of 10 students in my Math IIc class, and then buy them lunch, steak for the best, hamburger for the second best, and tacos for the third best. Then, this will a permutation, because "Order" matters by having different lunches with their rank,
thus $_{10}P_3 = \frac{10!}{(10-3)!} = \frac{10!}{7!}$

5) Combination is without a consideration of "ordering",
$_nC_r = \frac{n!}{(n-r)!r!}$
(eg) Suppose that I choose 3 students with the top best scores in my Math IIc class, and then buy them lunch, but this time, all hamburgers. Then, in this case, "order" does not matter, because they all eat hamburgers.

Thus, it is Combination, $_{10}C_3 = \frac{10!}{(10-3)!3!} = \frac{10!}{7!3!}$

Now, notice that $_{10}C_3$ and $_{10}P_3$ are exactly the same, except that $_{10}C_3$ is being divided by 3!, while $_{10}P_3$ isn't. That is because in $_{10}C_3$ case, those 3 students selected will have the same lunch, hamburger - hamburger - hamburger, or H - H - H $\rightarrow$ repetition of 3, therefore, being divided by 3!.

Based on the above ideas on Combination, we choose 4 cards out of 52 cards, but without any particular ordering.

Therefore, it will be combination, $_{52}C_4 = \frac{52!}{48!4!} = 270,725$.

Ans. (C)

89

18. The area of a triangle with sides 4, 5, and 8 is

    (A) 7.5
    (B) 8.18
    (C) 3.75
    (D) 13.0
    (E) 2.4

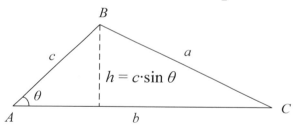

Here, we are given the lengths of 3 sides only, without any angle information.
Therefore, we use Heron's formula to find the area,
that is $s = \dfrac{a+b+c}{2} = \dfrac{4+5+8}{2} = 8.5$

Area $= \sqrt{S(s-a)(s-b)(s-c)}$
$= \sqrt{8.5(8.5-4)(8.5-5)(8.5-8)} = 8.18$

Ans. (B)

Now, in addition to Heron's formula, we also need to review following formulas for Math IIc - exam:

1) The Law of Sine: $\dfrac{\sin A}{a} = \dfrac{\sin B}{b} = \dfrac{\sin C}{c}$

2) The law of Cosine: $c^2 = a^2 + b^2 - 2ab\cdot\cos C$

3) Area with 2 side, and angle θ: Area $= \dfrac{1}{2} bc\cdot\sin\theta$

19. One side of a given triangle is 6 inches. Inside the triangle a line segment is drawn parallel to this side, cutting off a triangle whose area is one-thirds that of the given triangle. Find the length of this segment in inches.

    (A) 2
    (B) $2\sqrt{3}$
    (C) $3\sqrt{3}$
    (D) $6\sqrt{3}$
    (E) 9

Since the ratio of the area $\triangle DBE$ is $1:\dfrac{1}{3}$,

we get the ratio of the line segment $\overline{AC}:\overline{DE} = 1:\sqrt{\dfrac{1}{3}}$.

Ans. (B)

(Note)

Length : $1 \to 2 \to 3$
Area   : $1^2 \to 2^2 \to 3^2$
Volume : $1^3 \to 2^3 \to 3^3$

20. Given the statement, "Only if it snows will the class be cancelled", if it is known that it did not snow, what conclusion can be drawn?

(A) The class is cancelled.
(B) The class is not cancelled.
(C) The class might be cancelled.
(D) No conclusion can be drawn.
(E) None of the above.

As mentioned in the previous test #1, it is a logic problem with negation.

Ans. (B)

LOGIC :
There are 4 different logic cases;
Let's make a statement. If p, then q.
Using mathematical symbol for this statement is, $p \rightarrow q$.

- Negation: If p, then <u>NOT</u> q, or, $p \rightarrow \sim q$, where "$\sim$" sign represents "NOT".
- Inverse: If <u>NOT</u> p, then <u>NOT</u> q, or $\sim p \rightarrow \sim q$.
- Converse: If q, then p, or $q \rightarrow p$.
- Contra Positive: If <u>NOT</u> q, then <u>NOT</u> p, or $\sim q \rightarrow \sim p$.

Of all these, the Contra Positive is the most important one to remember.

Here, the Contra Positive is exactly the same statement as the original statement. Often, if we cannot prove the original statement, then we may use the logic of Contra Positive to prove the original statement.

21. What is the domain of the function $f(x) = \dfrac{\sqrt{2x^3+5}}{\sqrt{2x^3-5}}$?

(A) $x < -1.36$
(B) $-1.36 \leq x \leq 1.36$
(C) $x > 1.36$
(D) $x \geq 1.36$
(E) $x < -1.36$ or $x > 1.36$

In most case, a domain problem in Rational Function involves with the condition of its denominator.
In this problem, the denominator $\sqrt{2x^3 - 5} \neq 0$, and at the same time, $2x^3 - 5 > 0$.

$\therefore 2x^3 - 5 > 0$ or $x > \sqrt[3]{\dfrac{5}{2}}$

Ans. (C)

22. The hyperbola $9x^2 - 4y^2 = 16$ intersects the x-axis at approximately which of the following points?

(A) $(2, 0)$
(B) $(0, 2)$
(C) $(1.33, 0)$
(D) $(0, 1.33)$
(E) $(0, -1.33)$

This problem involves with the topics on Conic Sections. However, the question only asks about the x-intercept, when $y = 0$, we only need to plug in $y = 0$ into equation.
$9x^2 - 4(0)^2 = 16$

$\therefore x^2 = \dfrac{16}{9}$

$\therefore x = \pm\sqrt{\dfrac{16}{9}} = \pm\dfrac{4}{3} = \pm 1.33$

$\therefore (1.33, 0)$

Ans. (C)

(Note)

For the future reference, I want to go over topics on Conic Sections as follows:

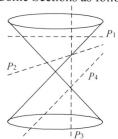

1) Circle: Let P_1 be the plane that cuts the cone in a horizontal direction. Then it creates a circle, formed by its intersection.

Given $(x - h)^2 + (y - k)^2 = r^2$,

- center = (h, k)
- radius = r

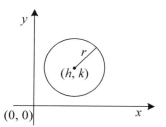

2) Ellipse: Let P_2 be the plane that cuts the cone in <u>a tilted direction</u>. Then it creates an elliptical figure, formed by intersection.

Given $\dfrac{(x-h)^2}{a^2} + \dfrac{(y-k)^2}{b^2} = 1$,

- center = (h, k)
- length of Major axis = $2a$, when $a > b$
- length of Minor axis = $2b$
- vertex = $(\pm a + h,\ o + k)$
- Foci = $(h \pm c, k)$, where $c^2 = a^2 - b^2$
- $d_1 + d_2 = 2a$

$$\frac{x^2}{a^2} + \frac{y^2}{b^2} = 1,\ a > b \qquad\qquad a < b$$

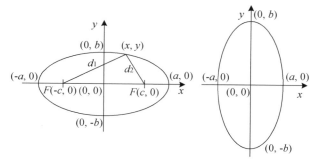

3) Hyperbola: Let P_3 be the plane that cuts the cone in a vertical direction. Then it creates a hyperbolic figure, formed by its intersection.

Given $\dfrac{(x-h)^2}{a^2} - \dfrac{(y-k)^2}{b^2} = 1$, or $\dfrac{(y-k)^2}{b^2} - \dfrac{(x-h)^2}{a^2} = 1$,

- center = (h, k)
- vertex = $(\pm a + h, o + k)$
- Asymptotes: $y - k = \pm \dfrac{b}{a}(x - h)$,

  where the slope $\dfrac{b}{a}$ is $\dfrac{\text{rise}}{\text{run}}$.
- Foci = $(\pm c + h, o + k)$, where $c^2 = a^2 + b^2$
- $d_1 - d_2 = 2a$

$$\frac{x^2}{a^2} - \frac{y^2}{b^2} = 1 \qquad\qquad \frac{y^2}{b^2} - \frac{x^2}{a^2} = 1$$

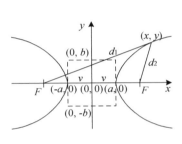
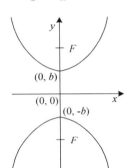

4) Parabola: Let P_4 be the plane that cuts the cone in a parallel to slant edge direction. Then it creates a parabolic figure, formed by its intersection.

Given $4P(y - k) = (x - h)^2$, or $4P(x - h) = (y - k)^2$,

- vertex = (h, k)
- Foci = $(o + h, p + k)$ or $(p + h, o + k)$
- Directrix line: $y = -p + k$ or $x = -p + h$

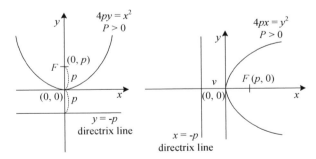

23. The length of the vector that could correctly be used to represent in the complex plane the number $-\sqrt{7} + 2i$ is

(A) 11 (B) $\sqrt{11}$ (C) $3\sqrt{2}$ (D) $\sqrt{5}$ (E) $\sqrt{13}$

The length of a vector, or the magnitude of a vector, $z = a + bi = \;<a, b>$ is given by $|z| = |a + bi| = \sqrt{a^2 + b^2}$.

Therefore, $|z| = |-\sqrt{7} + 2i| = \sqrt{(-\sqrt{7})^2 + 2^2} = \sqrt{11}$.

Ans. (B)

93

24. In a cube, the ratio of the longest diagonal to a side is

    (A) $\sqrt{2} : \sqrt{3}$
    (B) $\sqrt{6} : 2$
    (C) $\sqrt{6} : \sqrt{2}$
    (D) $\sqrt{3} : \sqrt{2}$
    (E) $\sqrt{3} : 1$

$c = \sqrt{1^2 + 1^2 + 1^2} = \sqrt{3}$
$b = \sqrt{1^2 + 1^2} = \sqrt{2}$
Thus, $c{:}a = \sqrt{3}{:}1$

Ans. (E)

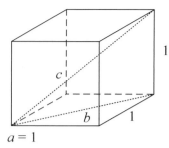

25. What is the measure of the smallest angle in a right triangle with sides of lengths 5, 12 and 13?

    (A) $21.04°$
    (B) $22.62°$
    (C) $24.62°$
    (D) $42.71°$
    (E) $67.34°$

Referring to the topics of Trig. Functions previously, we may use the Law of cosine, $c^2 = a^2 + b^2 - 2ab \cdot \cos C$.
Here, the shortest side is 5.
$\therefore 5^2 = 12^2 + 13^2 - 2 \cdot 12 \cdot 13 \cdot \cos C$.
$\therefore C = \cos^{-1} \dfrac{12^2 + 13^2 - 5^2}{2 \cdot 12 \cdot 13} = 22.62°$

Ans. (B)

26. What is the period of the curve whose equation is $y = 4\sin x \cdot \cos x$?

    (A) $60°$
    (B) $120°$
    (C) $180°$
    (D) $360°$
    (E) $720°$

$y = 4\sin x \cdot \cos x = 2(2\sin x \cdot \cos x)$.
But $2\sin x \cdot \cos x = \sin 2x$.
$\therefore y = 2 \cdot \sin 2x$
$\therefore p = 2 \dfrac{\pi}{2} = \pi = 180°$

Ans. (C)

27. Let $f(x)$ be a polynomial function.
If $f(\sqrt{2}) = 0$ and $f(-\sqrt{2}) = 0$, then $f(x)$ is divisible by

    (A) $x - 3$
    (B) $x^2 - 2$
    (C) $x^2 + 2$
    (D) $x^2 - 3x + 2$
    (E) $x^2 + 3x + 2$

Refer to the previous test #1, we have stated <u>Factor Theorem</u>, that is, if $f(x)$ has $f(a) = 0$, then it must have a factor of $(x - a)$. Here, $f(\sqrt{2}) = 0$ and $f(-\sqrt{2}) = 0$.
$\therefore$ we have $(x - \sqrt{2})(x + \sqrt{2})$ factors.
Therefore, $f(x)$ must be division by $(x^2 - 2)$.

Ans. (B)

28. How many real roots does the following equation have?
$5e^x + e^{-x} - 4 = 0$

    (A) 0
    (B) 1
    (C) 2
    (D) 4
    (E) an infinite number

Using a Graphing utility, let $y_1 = 5e^x + e^{-x} - 4$, and then check how many x-intercept it has.
There is no real root.

Ans. (A)

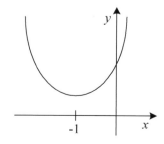

29. If $f(x) = \ln x$ and $g(x) = f(x) \cdot f^{-1}(x)$, what does $g(2)$ equal?

(A) 8.1   (B) 7.5   (C) 8.3   (D) 5.1   (E) 7.4

$f(x) = y = \ln x$.
Now, to get the inverse $f^{-1}(x)$ by switching x and y's,
$x = \ln y$ or $y = e^x = f^{-1}(x)$
$\therefore g(x) = f(x) \cdot f^{-1}(x) = (\ln x) \cdot (e^x)$
$\therefore g(2) = (\ln 2) \cdot (e^2) = 5.1$

Ans. (D)

30. If the 3rd term of a geometric progression is $\sqrt[3]{a}$ and the 6th term is $\sqrt[3]{a^2}$, what is the 12th term of the progression?

(A) a

(B) $2a$

(C) $a^{\frac{4}{3}}$

(D) $a^{\frac{5}{3}}$

(E) a^2

Recall the Geometric series with the following formulas:

$$a_n = a_1 r^{n-1}, \quad r = \frac{a^n}{a_{n-1}}, \quad s_n = a_1 \frac{1-r^n}{1-r}$$

And also, the infinite series, $s = a_1 \frac{1}{1-r}, \quad |r| < 1$.

Here, $a_3 = a^{\frac{1}{3}}$ and $a_6 = a^{\frac{2}{3}}$.

$\therefore \dfrac{a_6}{a_3} = \dfrac{a_1 r^5}{a_1 r^2} = r^3$, which is, $\dfrac{a^{\frac{2}{3}}}{a^{\frac{1}{3}}} = r^3$, or $a^{\frac{1}{3}} = r^3 \rightarrow r = a^{\frac{1}{9}}$.

Therefore, $a_{12} = a_6 \cdot r^6 = a^{\frac{2}{3}} \cdot (a^{\frac{1}{9}})^6 = a^{\frac{2}{3}} \cdot a^{\frac{2}{3}} = a^{\frac{4}{3}}$.

Ans. (C)

(Note)
For Arithmetic series,

$$a_n = a_1 + (n-1)d, \quad d = a_n - a_{n-1}, \quad s_n = \frac{a_1 + a_n}{2} \cdot n,$$
where d is the "Common Difference"

31. If $f(x, y, z) = \sqrt{3}x - \sqrt{5}y + \sqrt{6}z$ and $f(a, b, 0) = f(0, a, b)$, then $\dfrac{a}{b} =$

(A) 0.32   (B) 2.7   (C) 8.6   (D) 0.12   (E) 1.18

$f(x, y, z) = \sqrt{3}x - \sqrt{5}y + \sqrt{6}z$
$\therefore (a, b, o) = \sqrt{3}a - \sqrt{5}b + 0 \rightarrow$ eq. (1)
$\therefore (o, a, b) = 0 - \sqrt{5}a + \sqrt{6}b \rightarrow$ eq. (2)
$\therefore$ By equating (1) = (2), we get $\sqrt{3}a - \sqrt{5}b = -\sqrt{5}a + \sqrt{6}b$,
or $(\sqrt{3} + \sqrt{5})a = (\sqrt{6} + \sqrt{5})b \rightarrow \therefore \dfrac{a}{b} = \dfrac{\sqrt{6}+\sqrt{5}}{\sqrt{5}+\sqrt{3}} = 1.18$

Ans. (E)

32. What is the inverse of $f(x) = \dfrac{1}{\sqrt{-x^2+5}}$?

(A) $y = \dfrac{1}{\sqrt{5-x^2}}$

(B) $y = 5 - \dfrac{1}{x^2}$

(C) $y = \sqrt{5 - \dfrac{1}{x^2}}$

(D) $y = \sqrt{5 + \dfrac{1}{x^2}}$

(E) $y = \sqrt{-x^2 + 5}$

$y = f(x) = \dfrac{1}{\sqrt{-x^2+5}}$.
To get the inverse $f^{-1}(x)$ by switching x and y's,
we get $x = \dfrac{1}{\sqrt{-y^2+5}}$ or $x^2 = \dfrac{1}{-y^2+5}$.
Now, by reciprocal of both sides,
$\dfrac{1}{x^2} = -y^2 + 5 \rightarrow y^2 = 5 - \dfrac{1}{x^2}$, or $y = \sqrt{5 - \dfrac{1}{x^2}} = f^{-1}(x)$.

Ans. (C)

33. Find the value of P if $(x + 1)$ is a factor of $f(x) = 3x^4 + 2x^3 - Px^2 - 2x + 1$.

   (A) 1   (B) 2   (C) 3   (D) 4   (E) 5

If $(x + 1)$ is a factor of $f(x)$, then $f(-1) = 0$.
∴ $f(-1) = 3 - 2 - p + 2 + 1 = 0$.
∴ $p = 4$

Ans. (D)

34. In how many different orders can 10 students arrange themselves around a table?

   (A) 9   (B) 81   (C) 181,440
   (D) 362,880   (E) 387,420,489

Here, the circular permutation is is $(n - 1)!$
∴ $9! = 362,880$

Ans. (D)

35. If x and y are real numbers, which one of the following relations is a function of x?

   (A) $\{(x, y) \mid y = |x|\}$
   (B) $\{(x, y) \mid x = y^2 - 2y + 4\}$
   (C) $\{(x, y) \mid y = \pm\sqrt{1 - x}\}$
   (D) $\{(x, y) \mid y < x - 2\}$
   (E) $\{(x, y) \mid x = \sec y\}$

Vertical line testing shows $y = |x|$ is the only one that intersect with the function only once. Therefore, (A) is the only solution.

Ans. (A)

36. If the equation $3x^3 + 9x^2 + px - q = 0$ has 3 equal roots, then

   (A) $q = 0$
   (B) $p = 0$
   (C) $q = 1$
   (D) each root $= 1$
   (E) each root $= -1$

For a given cubic function,

$f(x) = ax^3 + bx^2 + cx + d$, the sum of the three roots,
$x_1 + x_2 + x_3 = (-)\dfrac{b}{a}$ and the product of the roots,
$x_1 \times x_2 \times x_3 = -\dfrac{d}{a}$

Here, the sum of 3 equal roots is $x_1 + x_2 + x_3$
$= x_1 + x_1 + x_1 = 3x_1 = (-)\dfrac{9}{3} = (-)3$.
∴ $x_1 = (-)1$

Ans. (E)

37. Each term of a sequence, after the first, is inversely proportional to the term preceding it. If the first two terms are 3 and 9, what is the 100th term?

   (A) 3
   (B) 9
   (C) 3^{100}
   (D) 3^{99}
   (E) The 100th term cannot be determined.

$a_1 = 3$, $a_2 = 9$, a_3, a_4, ...
Since the second term 9 is <u>inversely proportion</u> to the proceeding term 3, we get $9 = k \cdot \dfrac{1}{3}$,
∴ $k = 27$.

Therefore the general term $a_n = k\dfrac{1}{a_{n-1}} = 27\dfrac{1}{a_{n-1}}$.
Using this sequence, $a_3 = 27\dfrac{1}{a_2} = 27\dfrac{1}{9} = 3$,
$a_4 = 27\dfrac{1}{a_3} = 27\dfrac{1}{3} = 9$, ... etc., which creates the sequence of <u>3</u>, <u>9</u>, <u>3</u>, <u>9</u>, ... Every even term is 9.
∴ $a_{100} = 9$

Ans. (B)

38. The point whose polar coordinates are $(3, 45°)$ is the same as the point whose polar coordinates are

(A) $(-3, 45°)$
(B) $(-3, 225°)$
(C) $(3, -45°)$
(D) $(-3, 135°)$
(E) $(3, 225°)$

As shown in the figure,
the point Q is in the opposite side of the point P.
Now, for the point $Q(3, 225°)$ to be the same point as $P(3, 45°)$, we need to change $r = (+)3$ of point Q to $r = (-)3$ to move the direction backward!!
∴ $(-3, 225°) = (3, 45°)$
In general, $(r, \theta) = (-r, \theta + 180°)$.

Ans. (B)

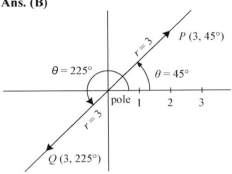

39. If $\log x^2 \geq \log 8 + \dfrac{1}{2} \log x$, then

(A) $x \geq 2$
(B) $x \leq 2$
(C) $x \leq 4$
(D) $x \geq 4$
(E) $x \geq 1$

Referring to <u>the properties of logarithmic function,</u>
$\log AB = \log A + \log B$, $\log \dfrac{A}{B} = \log A - \log B$,

we get $\log x^2 - \log 8 - \log \sqrt{x} \geq 0$, or $\dfrac{x^2}{8\sqrt{x}} \geq 1 \rightarrow$

$\dfrac{x^2}{8\sqrt{x}} \geq 1 \rightarrow x^2 \geq 8\sqrt{x} \rightarrow x^2 - 8\sqrt{x} \geq 0 \rightarrow \sqrt{x} \cdot (\sqrt{x^3} - 8) \geq 0.$
Since $\sqrt{x} \geq 0$, we get $\sqrt{x^3} - 8 \geq 0.$
∴ $\sqrt{x^3} \geq 8$ ∴ $x \geq 4.$

Ans. (D)

However, this algebraic manipulation takes too long. Therefore, my recommendation is to use a graphing utility, let $y_1 = \log x^2$, and $y_2 = \log 8 + \dfrac{1}{2} \log x$, and then, look for the solution for $y_1 \geq y_2 \rightarrow x \geq 4.$

40. A right circular cylinder is circumscribed about a sphere. If V_s represents the volume of the sphere and V_c represents the volume of the cylinder, then

(A) $V_s = \dfrac{2}{3} V_c$

(B) $V_s < \dfrac{2}{3} V_c$

(C) $V_s > \dfrac{2}{3} V_c$

(D) $V_s \leq \dfrac{2}{3} V_c$

(E) $V_s \geq \dfrac{2}{3} V_c$

Refer to the figure shown,
$V_c = \pi r^2 h = \pi r^2 (2r) = 2\pi r^3.$
$V_s = \dfrac{4}{3} \pi r^3$
∴ $V_s = \dfrac{2}{3} V_c$

Ans. (A)

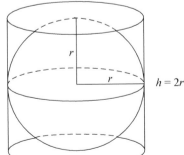

97

41. $f(x) = \begin{cases} \dfrac{x^2-2x-3}{x-3}, & \text{when } x \neq 3 \\ k, & \text{when } x = 3 \end{cases}$

what must the value of k be in order for $f(x)$ to be a continuous function?

(A) 2
(B) 4
(C) 0
(D) -2
(E) No value of k will make $f(x)$ a continuous function.

In order to be a continuous function, the discontinuity at $x = 3$ must be removed. That means, for the given function,

$f(x) = \dfrac{x^2-2x-3}{x-3} = \dfrac{(x-3)(x+1)}{(x-3)}$ must be

reduced to $f(x) = \dfrac{(x-3)(x+1)}{(x-3)} = (x+1), x \neq 3.$

Now, with this function $f(x) = (x+1), f(3) = 3 + 1 = 4,$ and k must be equal to $f(3)$ for $f(x)$ to be a continuous function.

Ans. (B)

42. Write the complex number $\sqrt{3}$ - i in polar form.

(A) $2(\cos\dfrac{\pi}{6} + i\cdot\sin\dfrac{\pi}{6})$

(B) $-2(\cos\dfrac{\pi}{6} + i\cdot\sin\dfrac{\pi}{6})$

(C) $2(\cos\dfrac{7\pi}{6} + i\cdot\sin\dfrac{7\pi}{6})$

(D) $-2(\cos\dfrac{11\pi}{6} + i\cdot\sin\dfrac{11\pi}{6})$

(E) $2(\cos\dfrac{11\pi}{6} + i\cdot\sin\dfrac{11\pi}{6})$

Let $z = a + bi = \sqrt{3}$ - i, where $a = \sqrt{3}, b = (-)1.$
Then, from the course of <u>Polar coordinates</u>,
$r = \sqrt{a^2 + b^2} = \sqrt{4} = 2, \theta = \tan^{-1}(-\dfrac{1}{\sqrt{3}}) = -\dfrac{\pi}{6},$

or $\dfrac{11\pi}{6}$ in the 4th quadrant.

Therefore, using $z = a + bi = r(\cos\theta + i\sin\theta)$ in polar form,

$z = \sqrt{3}$ - $i = 2(\cos\dfrac{11\pi}{6} + i\sin\dfrac{11\pi}{6}).$

Ans. (E)

43. If $\dfrac{\cos^2\theta - \sin^2\theta}{\sin\theta\cos\theta} = 2\sqrt{3}$, then $\theta =$

(A) 15°   (B) 30°   (C) 45°   (D) 60°   (E) 75°

Using a graphing utility, let $y_1 = \dfrac{\cos 2\theta - \sin 2\theta}{\sin\theta\cos\theta}$,

$y_2 = 2\sqrt{3}$, and then find the intersection.
The result is $\theta = 15°.$

Ans. (A)

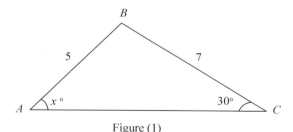

Figure (1)

44. What is the degree measure of the angle x of a triangle that has sides of length of 5 and 7 as shown in the figure (1)?

(A) 44.43°
(B) 54.31°
(C) 71.37°
(D) 25.69°
(E) 34.31°

Using the Law of sine, we get $\dfrac{5}{\sin 30°} = \dfrac{7}{\sin x°}$

$\therefore \sin x = \dfrac{7\sin 30°}{5} = \dfrac{7}{10}$

$\therefore x = \sin^{-1}\dfrac{7}{10} = 44.43°$

Ans. (A)

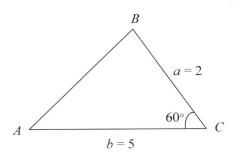

45. In $\triangle ABC$, $a = 2$, $b = 5$, $\angle C = 60°$.
    What is the area of $\triangle ABC$?

    (A) 4.3
    (B) 3.6
    (C) 3.2
    (D) 2.9
    (E) 2.3

Area of $\triangle ABC = \dfrac{1}{2} \cdot a \cdot b \cdot \sin c° = \dfrac{1}{2}(2)(5) \cdot \sin 60° = \dfrac{5\sqrt{3}}{2}$.

Ans. (A)

46. If $f(x, y) = x^3 - xy^2 + 3y^4$,
    which of the following is (are) true?

    I. $f(x, y) = f(x, -y)$
    II. $f(x, y) = f(-x, y)$
    III. $f(x, y) = f(-x, -y)$

    (A) I only
    (B) II only
    (C) III only
    (D) I and III only
    (E) I, II, and III

Since $f(x, y) = x^3 - xy^2 + 3y^4$ has y–terms with even powers, $f(x, -y)$ should have the same result as $f(x, y)$, while $f(-x, y)$ should have effect on its x–terms with odd powers. Therefore, the only right choice is (I).

Ans. (A)

47. Figure 5 shows a cube with edge of length 4 centimeters.
    If points A and C are midpoints of the edges of the cube, what is the area of region $ABCD$?

$\overline{AD} = \sqrt{2^2 + 4^2} = \sqrt{20} = 2\sqrt{5}$.
$\therefore$ Area of $\square ABCD = \left(2\sqrt{5}\right)^2 = 20\text{cm}^2$

Ans. (D)

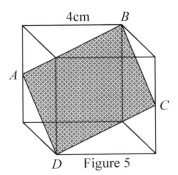

D   Figure 5

    (A) 15.71 cm²
    (B) 17.25 cm²
    (C) 19.60 cm²
    (D) 20.00 cm²
    (E) 22.63 cm²

48. If (x, y) represents a point on the graph of $y = \sqrt{2x + 1}$, which of the following could be a portion of the graph of the set of points (x, y^2)?

(A)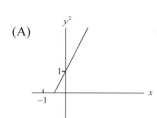

(B)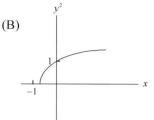

(C)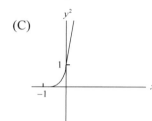

(D)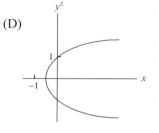

(E)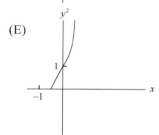

For a given function, $y = \sqrt{2x + 1}$, we transform our coordinates from (x, y) to (x, y^2). Therefore, letting $Y = y^2$, we get $Y = y^2 = \left(\sqrt{2x + 1}\right)^2 = 2x + 1$, or $Y = 2x + 1$ in (x, Y) coordinate system.

Ans. (A)

49. If the parameter is eliminated from the equations $x = 4t^2 + 1$ and $y = 2t$, then the relation between x and y is

(A) $y = x - 1$
(B) $y = 1 - x$
(C) $y^2 = x - 1$
(D) $y^2 = (x - 1)^2$
(E) $y^2 = 4x - 4$

In this problem of parametric equation, we need to eliminate "t" from x and y equations, and set up the two equation into one equation with x and y variables only.
Since $y = 2t$, and $x = 4t^2 + 1$, we replace $4t^2$ with y^2.
$\therefore x = 4t^2 + 1 = (2t)^2 + 1 = y^2 + 1$
$\therefore y^2 = x - 1$

Ans. (C)

50. If 8 people shook hands with each other, how many handshakes were exchanged?

(A) 8
(B) 16
(C) 21
(D) 28
(E) 56

Shaking hands implies between 2 people.
Therefore, $_8C_2 = 28$.

Ans. (D)

Model Test No. 03

50 Questions / 60 Minutes

Directions: For each question, determine which of the answer choices is correct and fill in the oval on the answer sheet that corresponds to your choice.

Notes:

1. You will need to use a scientific or graphing calculator to answer some of the questions.
2. Be sure your calculator is in degree mode.
3. Each figure on this test is drawn as accurately as possible unless it is specifically indicated that the figure has not been drawn to scale.
4. The domain of any function f is the set of all real numbers x for which $f(x)$ is also a real number, unless the question indicates that the domain has been restricted in some way.
5. The box below contains five formulas that you may need to answer one or more of the questions.

REFERENCE INFORMATION

THE FOLLOWING INFORMATION IS FOR YOUR REFERENCE IN ANSWERING SOME OF THE QUESTIONS IN THIS TEST.

Volume of a right circular cone with radius r and height h: $V = \dfrac{1}{3}\pi r^2 h$

Lateral Area of a right circular cone with circumference of the base c and slant height l: $S = \dfrac{1}{2}cl$

Volume of a sphere with radius r: $V = \dfrac{4}{3}\pi r^3$

Surface Area of a sphere with radius r: $S = 4\pi r^2$

Volume of a pyramid with base area B and height h: $V = \dfrac{1}{3}Bh$

Answer Sheet
Model Test No. 03

1 Ⓐ Ⓑ Ⓒ Ⓓ Ⓔ	14 Ⓐ Ⓑ Ⓒ Ⓓ Ⓔ	27 Ⓐ Ⓑ Ⓒ Ⓓ Ⓔ	40 Ⓐ Ⓑ Ⓒ Ⓓ Ⓔ
2 Ⓐ Ⓑ Ⓒ Ⓓ Ⓔ	15 Ⓐ Ⓑ Ⓒ Ⓓ Ⓔ	28 Ⓐ Ⓑ Ⓒ Ⓓ Ⓔ	41 Ⓐ Ⓑ Ⓒ Ⓓ Ⓔ
3 Ⓐ Ⓑ Ⓒ Ⓓ Ⓔ	16 Ⓐ Ⓑ Ⓒ Ⓓ Ⓔ	29 Ⓐ Ⓑ Ⓒ Ⓓ Ⓔ	42 Ⓐ Ⓑ Ⓒ Ⓓ Ⓔ
4 Ⓐ Ⓑ Ⓒ Ⓓ Ⓔ	17 Ⓐ Ⓑ Ⓒ Ⓓ Ⓔ	30 Ⓐ Ⓑ Ⓒ Ⓓ Ⓔ	43 Ⓐ Ⓑ Ⓒ Ⓓ Ⓔ
5 Ⓐ Ⓑ Ⓒ Ⓓ Ⓔ	18 Ⓐ Ⓑ Ⓒ Ⓓ Ⓔ	31 Ⓐ Ⓑ Ⓒ Ⓓ Ⓔ	44 Ⓐ Ⓑ Ⓒ Ⓓ Ⓔ
6 Ⓐ Ⓑ Ⓒ Ⓓ Ⓔ	19 Ⓐ Ⓑ Ⓒ Ⓓ Ⓔ	32 Ⓐ Ⓑ Ⓒ Ⓓ Ⓔ	45 Ⓐ Ⓑ Ⓒ Ⓓ Ⓔ
7 Ⓐ Ⓑ Ⓒ Ⓓ Ⓔ	20 Ⓐ Ⓑ Ⓒ Ⓓ Ⓔ	33 Ⓐ Ⓑ Ⓒ Ⓓ Ⓔ	46 Ⓐ Ⓑ Ⓒ Ⓓ Ⓔ
8 Ⓐ Ⓑ Ⓒ Ⓓ Ⓔ	21 Ⓐ Ⓑ Ⓒ Ⓓ Ⓔ	34 Ⓐ Ⓑ Ⓒ Ⓓ Ⓔ	47 Ⓐ Ⓑ Ⓒ Ⓓ Ⓔ
9 Ⓐ Ⓑ Ⓒ Ⓓ Ⓔ	22 Ⓐ Ⓑ Ⓒ Ⓓ Ⓔ	35 Ⓐ Ⓑ Ⓒ Ⓓ Ⓔ	48 Ⓐ Ⓑ Ⓒ Ⓓ Ⓔ
10 Ⓐ Ⓑ Ⓒ Ⓓ Ⓔ	23 Ⓐ Ⓑ Ⓒ Ⓓ Ⓔ	36 Ⓐ Ⓑ Ⓒ Ⓓ Ⓔ	49 Ⓐ Ⓑ Ⓒ Ⓓ Ⓔ
11 Ⓐ Ⓑ Ⓒ Ⓓ Ⓔ	24 Ⓐ Ⓑ Ⓒ Ⓓ Ⓔ	37 Ⓐ Ⓑ Ⓒ Ⓓ Ⓔ	50 Ⓐ Ⓑ Ⓒ Ⓓ Ⓔ
12 Ⓐ Ⓑ Ⓒ Ⓓ Ⓔ	25 Ⓐ Ⓑ Ⓒ Ⓓ Ⓔ	38 Ⓐ Ⓑ Ⓒ Ⓓ Ⓔ	
13 Ⓐ Ⓑ Ⓒ Ⓓ Ⓔ	26 Ⓐ Ⓑ Ⓒ Ⓓ Ⓔ	39 Ⓐ Ⓑ Ⓒ Ⓓ Ⓔ	

1. If the graph of $3x + y + 5 = 0$ is perpendicular to the graph of $ax - 2y + 9 = 0$, then a equals

   (A) $\dfrac{3}{2}$

   (B) $-\dfrac{3}{2}$

   (C) 6

   (D) -6

   (E) $\dfrac{2}{3}$

2. Which of the following is equivalent to $a - b > a + b$?

   (A) $a > b$
   (B) $a > 0$
   (C) $b > a$
   (D) $b > 0$
   (E) $b < 0$

3. If $f(x) = ax^2 + bx + c$ for all real numbers x and if $f(0) = 3$ and $f(1) = 4$, then $a + b =$

   (A) -2
   (B) -1
   (C) 0
   (D) 1
   (E) 2

4. If $\sqrt{3x} = 3.35$, then $x =$

   (A) 0.62
   (B) 1.93
   (C) 3.74
   (D) 5.33
   (E) 11.59

5. If $f(r, \theta) = r \sin \theta$, then $f(3, -1) =$

   (A) -2.52
   (B) -1.98
   (C) 0.10
   (D) 2.00
   (E) 1.25

105

6. The average of your first five test grades is 75. What grade must you get on your sixth test to make your average 70?

(A) 80
(B) 82
(C) 84
(D) 45
(E) 88

7. If $\log_a 9 = 2$, then what is a?

(A) 0.89
(B) 0.95
(C) 3
(D) 1.21
(E) 7.61

8. If $\dfrac{3x-y}{y} = 2$, what is the value of $\dfrac{x}{y}$?

(A) $-\dfrac{8}{3}$

(B) -2
(C) 1

(D) $\dfrac{8}{3}$

(E) 2

9. If $\cos\theta = \dfrac{2}{3}$, then $\sin\theta =$

(A) $\dfrac{2\sqrt{13}}{13}$

(B) $\pm\dfrac{\sqrt{5}}{3}$

(C) $\dfrac{3\sqrt{13}}{13}$

(D) $\pm\dfrac{2}{5}$

(E) $\dfrac{2\sqrt{5}}{5}$

10. Approximate $\log_8 13$

(A) 1.23
(B) 1.21
(C) 1.18
(D) 1.15
(E) 1.02

11. The circumference of circle $x^2 + y^2 - 6x - 4y - 36 = 0$ is

(A) 192
(B) 38
(C) 125
(D) 54
(E) 44

USE THIS SPACE FOR SCRATCH WORK

12. If $f(x) = 4$ for all real numbers x, then $f(x - 2) =$

(A) 0
(B) 2
(C) 4
(D) x
(E) The value cannot be determined.

13. In right triangle ABC, $AB = 13$, $BC = 12$, $AC = 5$. The tangent of $\angle A$ is

(A) $\dfrac{12}{13}$

(B) $\dfrac{13}{12}$

(C) $\dfrac{12}{5}$

(D) $\dfrac{5}{12}$

(E) $\dfrac{5}{13}$

14. The sum of the roots of the equation $(x + \sqrt{2})^2(2x - \sqrt{3})(x - \sqrt{7}) = 0$ is

(A) 3.3
(B) 0.7
(C) 2.1
(D) 6.3
(E) 2.5

15. What is the length of the major axis of the ellipse whose equation is $10x^2 + 6y^2 = 15$?

(A) 1.26
(B) 2.50
(C) 3.16
(D) 4.47
(E) 5.00

16. If $f(g(x)) = \dfrac{\frac{3}{2}\sqrt{x^2-1}-1}{2\sqrt{x^2-1}+1}$ and $f(x) = \dfrac{\frac{3}{2}x-1}{2x+1}$, then $g(x) =$

(A) $\sqrt{x}$
(B) $\sqrt{x^2-1}$
(C) x
(D) x^2
(E) x^2+1

17. If A is the degree measure of an acute angle and $\sin A = 0.25$, then $\sin A \cdot \cos(90° - A) =$

(A) 0.02
(B) 0.04
(C) 0.05
(D) 0.06
(E) 0.08

18. The plane $x + 2y + 3z = 5$ intersects the x-axis at $(a,0,0)$, the y-axis at $(0,b,0)$, and the z-axis at $(0,0,c)$.
What is the volume of the pyramid in the first octant with vertices, a, b, c and the origin?

(A) 5
(B) $\dfrac{125}{36}$
(C) $\dfrac{65}{12}$
(D) 1
(E) 9

19. If $f(x,y) = x^2 - 2y^2$ and $g(x) = 3^x$,
the value of $g(f(2,1)) =$

(A) 1
(B) 4
(C) 9
(D) -4
(E) 0

20. The units digit of 2567^{93} is

(A) 1
(B) 3
(C) 7
(D) 9
(E) none of the above

108

21. Solve for r: $5^{7-r} = 25^{r-1}$

(A) 1
(B) 2
(C) 3
(D) 4
(E) 5

22. What is $\lim\limits_{x \to \infty} \dfrac{4x^2 - \sqrt{x} + 5}{3x^2 + \sqrt{7}}$?

(A) -1.12
(B) .91
(C) 1.11
(D) 1.33
(E) 1.50

23. A rod, pivoted at one end, rotates through $\dfrac{\pi}{4}$ radians. If the rod is 16 inches long, how many inches does the free end travel?

(A) π
(B) 2π
(C) 3π
(D) 4π
(E) $\dfrac{3\pi}{2}$

24. The graph of $y = 2x - |x|$ is equivalent to the graph of

(A) $y = x$ for $x \leq 0$
(B) $y = x$ for $x \geq 0$
(C) $y = 2x$ for $0 \leq x \leq 1$
(D) $y = 2x$ for $x \leq 0$
(E) $y = 2x$ for $x \geq 0$

25. If the radius of a sphere is doubled, the percent increase in surface area is

(A) 100
(B) 300
(C) 400
(D) 700
(E) 800

26. Find the value, in simplest form, of $\dfrac{(2^{2n})(2^{n+1})-(2^n)^3}{2(2^{3n})}$

(A) $\dfrac{1}{2}$

(B) $\dfrac{1}{4}$

(C) $\dfrac{3}{4}$

(D) $\dfrac{5}{8}$

(E) $\dfrac{7}{8}$

27. If $\log_9 p = 3$ and $\log_3 q = 4$, p expressed in terms of q is

(A) $p = q^2$
(B) $p = q^8$
(C) $p = q^{\frac{3}{2}}$
(D) $p = q^{\frac{5}{2}}$
(E) $p = q^3$

28. If $a = 3\sec\theta$, and $b = 3\tan\theta$, then $\sqrt{a^2 - b^2}$

(A) 1
(B) 3
(C) 6
(D) $3\cos\theta\cdot\sin\theta$
(E) $3(\cos\theta + \sin\theta)$

29. The value of $\dfrac{525!}{521!\,3!}$ is

(A) greater than 10^{100}
(B) between 10^{10} and 10^{100}
(C) between 10^5 and 10^{10}
(D) between 10 and 10^5
(E) less than 10

30. If $f(x) = ax + b$, which of the following make(s) $f(x) = f^{-1}(x)$?

I. $a = -1$, $b =$ any real number
II. $a = 1$, $b = 0$
III. $a =$ any real number, $b = 0$

(A) only I
(B) only II
(C) only III
(D) only I and II
(E) only I and III

31. If the roots of the equation $ax^2 + bx + c = 0$ are α and β, the value of $\dfrac{(\alpha+\beta)^2}{\alpha\beta}$ in terms of a, b and c is

(A) $\dfrac{b^2}{a^2c^2}$

(B) $\dfrac{a^2c^2}{b}$

(C) $\dfrac{b^2}{ac}$

(D) $\dfrac{-b}{a^2c^2}$

(E) $\dfrac{-b^2}{ac}$

32. The range of the function $y = -x^{-\frac{2}{3}}$ is

(A) $y < 0$
(B) $y > 0$
(C) $y \geq 0$
(D) $y \leq 0$
(E) all real numbers

33. If P represents the set of rhombi and Q the set of squares, then the set $P \cap Q$ represents the set of

(A) squares
(B) trapezoids
(C) parallelograms
(D) quadrilaterals
(E) rectangles

111

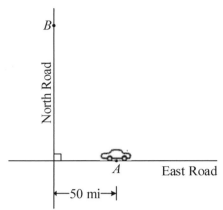

North Road

B•

A
East Road

|←50 mi→|

34. The figure above shows a car traveling from point A to point B. The straight-line distance between points A and B is 130 miles. If the car travels at an average speed of 65 miles per hour along East and North Roads, how long will it take the car to get to point B?

(A) 51 minutes
(B) 2 hour and 7 minutes
(C) 2 hour and 37 minutes
(D) 2 hour and 33 minutes
(E) 2 hour and 46 minutes

35. If y varies jointly as x, w, and the cube of z, what is the effect on w when x, y, and z are doubled?

(A) w is doubled
(B) w is multiplied by 4
(C) w is multiplied by 8
(D) w is divided by 8
(E) w is divided by 4

36. $\sum_{k=1}^{4} 2\left(\frac{2}{3}\right)^{k} =$

(A) $3\dfrac{17}{81}$

(B) $4\dfrac{20}{27}$

(C) $5\dfrac{20}{27}$

(D) $4\dfrac{1}{3}$

(E) $5\dfrac{7}{27}$

37. $F(x) = \dfrac{5x^2 - 5}{x - 1}$, when $x \neq 1$

      $= k$, when $x = 1$

For what value(s) of k is F a continuous function?

(A) 1
(B) 2
(C) 3
(D) 10
(E) no value of k

38. What is the smallest positive x-intercept of
the graph of $y = \dfrac{1}{2} \sin (4x + \dfrac{4\pi}{3})$?

(A) 0
(B) 0.52
(C) 1.05
(D) 2.09
(E) 1.31

39. How many integers greater than 2000 can be formed
from the digits 1, 2, 3, and 4, if no digit is repeated
in any number?

(A) 9
(B) 18
(C) 27
(D) 36
(E) 72

40. If $x(t) = 5\sin t$, $y(t) = 3 + 4\tan t$,
what is the approximate value of x when $y = 6$?

(A) 1
(B) 3
(C) 4
(D) 5
(E) 6

41. If the roots of the equation $x^3 + px^2 + qx + r = 0$ are
s_1, s_2 and s_3, then $s_1{}^2 + s_2{}^2 + s_3{}^2 =$

(A) $p^2 + q^2$
(B) $p^2 - 2q$
(C) $p^2 - q^2$
(D) p^2
(E) q^2

42. Find the area of the regular hexagon inscribed in a circle of radius 6.

(A) 36
(B) $36\sqrt{3}$
(C) 54
(D) $54\sqrt{3}$
(E) $128\sqrt{3}$

43. What is the period of $y = 5 - \cos 4x$?

(A) $\dfrac{\pi}{4}$

(B) $\dfrac{\pi}{2}$

(C) π

(D) $\dfrac{3\pi}{2}$

(E) 2π

44. For what value of x is the function $\cos^2 x - \sin^2 x$ a minimum?

(A) 2π

(B) $\dfrac{3\pi}{4}$

(C) $\dfrac{\pi}{2}$

(D) $\dfrac{2\pi}{3}$

(E) $\dfrac{\pi}{4}$

45. What is the equation of the perpendicular bisector of the line segment whose end points are (-3,5) and (5,-1)?

(A) $4x - 3y + 2 = 0$
(B) $4x + 3y - 13 = 0$
(C) $x - 3y + 10 = 0$
(D) $x + 3y - 8 = 0$
(E) $4x + 3y - 1 = 3$

46. If $3(x + 1)^2 + 10(y - 2)^2 = 48$ is graphed the sum of the distances from any point on the curve to the two foci is

(A) 4
(B) 8
(C) 12
(D) 16
(E) 32

114

47. Which of the following is the equation of the circle that has its center at $(1, 1)$ and is tangent to the line with equation $x + 3y = -6$?

(A) $(x-1)^2 + (y-1)^2 = 2$
(B) $(x-1)^2 + (y-1)^2 = 4$
(C) $(x-1)^2 + (y-1)^2 = 3$
(D) $(x-1)^2 + (y-1)^2 = 5$
(E) $(x-1)^2 + (y-1)^2 = 10$

48. The polar coordinates of a point P are $(2, \frac{4\pi}{3})$. The Cartesian (rectangular) coordinates of P are

(A) $(-1, -\sqrt{3})$
(B) $(-1, \sqrt{3})$
(C) $(-\sqrt{3}, -1)$
(D) $(-\sqrt{3}, 1)$
(E) none of the above

49. If the magnitudes of vectors $\mathbf{u}$ and $\mathbf{v}$ are 4 and 8, respectively, then the magnitude of vector $(\mathbf{v} - \mathbf{u})$ could NOT be

(A) 5
(B) 7
(C) 10
(D) 12
(E) 3

50. Given the statement "All boys play baseball," which of the following negates this statement?

(A) All boys play tennis.
(B) Some girls play tennis.
(C) All boys do not play tennis.
(D) At least one boy doesn't play baseball.
(E) All girls do not play tennis.

USE THIS SPACE FOR SCRATCH WORK

115

Answer Key
Model Test No. 03

#	Ans	#	Ans	#	Ans	#	Ans
1	E	14	B	27	C	40	B
2	E	15	C	28	B	41	B
3	D	16	B	29	B	42	D
4	C	17	D	30	D	43	B
5	A	18	B	31	C	44	C
6	D	19	C	32	A	45	A
7	C	20	C	33	A	46	B
8	C	21	C	34	C	47	E
9	B	22	D	35	D	48	A
10	A	23	D	36	A	49	E
11	E	24	B	37	D	50	D
12	C	25	B	38	B		
13	C	26	A	39	B		

How to Score the SAT Subject Test in Mathematics Level 2

When you take an actual SAT Subject Test in Mathematics Level 2, your answer sheet will be "read" by a scanning machine that will record your responses to each question. Then a computer will compare your answers with the correct answers and produce your raw score. You get one point for each correct answer. For each wrong answer, you lose one-fourth of a point. Questions you omit (and any for which you mark more than one answer) are not counted. This raw score is converted to a scaled score that is reported to you and to the colleges you specify.

Finding Your Raw Test Score

STEP 1: Table A lists the correct answers for all the questions on the Subject Test in Mathematics Level 2 that is reproduced in this book. It also serves as a worksheet for you to calculate your raw score.

- Compare your answers with those given in the table.
- Put a check in the column marked "Right" if your answer is correct.
- Put a check in the column marked "Wrong" if your answer is incorrect.
- Leave both columns blank if you omitted the question.

STEP 2: Count the number of right answers.
Enter the total here: _____

STEP 3: Count the number of wrong answers.
Enter the total here: _____

STEP 4: Multiply the number of wrong answers by .250.
Enter the product here: _____

STEP 5: Subtract the result obtained in Step 4 from the total you obtained in Step 2.
Enter the result here: _____

STEP 6: Round the number obtained in Step 5 to the nearest whole number.
Enter the result here: _____

The number you obtained in Step 6 is your raw score.

Scaled Score Conversion Table
Subject Test in Mathematics Level 2

Raw Score	Scaled Score	Raw Score	Scaled Score	Raw Score	Scaled Score
50	800	28	630	6	470
49	800	27	630	5	460
48	800	26	620	4	450
47	800	25	610	3	440
46	800	24	600	2	430
45	800	23	600	1	420
44	800	22	590	0	410
43	790	21	580	-1	400
42	780	20	580	-2	390
41	770	19	570	-3	370
40	760	18	560	-4	360
39	750	17	560	-5	350
38	740	16	550	-6	340
37	730	15	540	-7	340
36	710	14	530	-8	330
35	700	13	530	-9	330
34	690	12	520	-10	320
33	680	11	510	-11	310
32	670	10	500	-12	300
31	660	9	490		
30	650	8	480		
29	640	7	480		

1. If the graph of $3x + y + 5 = 0$ is perpendicular to the graph of $ax - 2y + 9 = 0$, then a equals

    (A) $\frac{3}{2}$

    (B) $-\frac{3}{2}$

    (C) 6

    (D) -6

    (E) $\frac{2}{3}$

    To be perpendicular to each other, the product of the slopes, $m_1 m_2 = -1$, we have $y_1 = -3x - 5$, $y_2 = \frac{a}{2} x + \frac{9}{2}$

    $\therefore (-3)\frac{a}{2} = -1$

    $\therefore a = \frac{2}{3}$

    Ans. (E)

2. Which of the following is equivalent to $a - b > a + b$?

    (A) $a > b$ (B) $a > 0$ (C) $b > a$
    (D) $b > 0$ (E) $b < 0$

    $a - b > a + b \rightarrow$ by adding $(-)a$ on both sides, we get $-b > b$, or $0 > 2b$.
    $\therefore b < 0$

    Ans. (E)

3. If $f(x) = ax^2 + bx + c$ for all real numbers x and if $f(0) = 3$ and $f(1) = 4$, then $a + b =$

    (A) -2   (B) -1   (C) 0   (D) 1   (E) 2

    $f(0) = c = 3$
    $\therefore c = 3$ $f(1) = a + b + c = 4$
    But plug in $c = 3$, we get $a + b + 3 = 4$.
    $\therefore a + b = 1$

    Ans. (D)

4. If $\sqrt{3x} = 3.35$, then $x =$

    (A) 0.62   (B) 1.93   (C) 3.74
    (D) 5.33   (E) 11.59

    Squaring both sides, we get $\left(\sqrt{3x}\right)^2 = (3.35)^2$
    $\therefore 3x = 11.225$
    $\therefore x = 3.74$

    Ans. (C)

5. If $f(r, \theta) = r \sin \theta$, then $f(3, -1) =$

    (A) -2.52   (B) -1.98   (C) 0.10
    (D) 2.00   (E) 1.25

    Plugging in $r = 3$, $\theta = -1$ in radian mode, we get $f(3, -1) = 3\sin(-1)$

    Ans. (A)

6. The average of your first five test grades is 75. What grade must you get on your sixth test to make your average 70?

    (A) 80   (B) 82   (C) 84   (D) 45   (E) 88

    The sum of the first five tests is $75 \times 5 = 375$.
    Now, let $x =$ the grade of the sixth test.
    Then, the average of the six tests will be, $\frac{375+x}{6} = 70$.
    $\therefore x = 420 - 375 = 45$

    Ans. (D)

7. If $\log_a 9 = 2$, then what is a?

    (A) 0.89   (B) 0.95   (C) 3   (D) 1.21   (E) 7.61

    $\log_a 9 = 2 \rightarrow 9 = a^2$ by the definition of logarithm.
    $\therefore a = 3$

    Ans. (C)

8. If $\dfrac{3x-y}{y} = 2$, what is the value of $\dfrac{x}{y}$?

   (A) $-\dfrac{8}{3}$

   (B) -2

   (C) 1

   (D) $\dfrac{8}{3}$

   (E) 2

$\dfrac{3x-y}{y} = 2 \rightarrow 3\dfrac{x}{y} - \dfrac{y}{y} = 3\dfrac{x}{y} - 1 = 2$

$\therefore 3\dfrac{x}{y} = 3$

$\therefore \dfrac{x}{y} = 1$

Ans. (C)

9. If $\cos\theta = \dfrac{2}{3}$, then $\sin\theta =$

   (A) $\dfrac{2\sqrt{13}}{13}$

   (B) $\pm\dfrac{\sqrt{5}}{3}$

   (C) $\dfrac{3\sqrt{13}}{13}$

   (D) $\pm\dfrac{2}{5}$

   (E) $\dfrac{2\sqrt{5}}{5}$

Using calculator for $\theta = \cos^{-1}\dfrac{2}{3}$, we get $\theta = 48.2°$.

$\therefore \sin(48.2°) = 0.745$ or $\sin\theta = \dfrac{\sqrt{5}}{3}$.

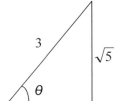

But $\cos\theta = \dfrac{2}{3} > 0$.

θ must be in the first or the fourth quadrant!!

$\therefore \sin\theta = \pm\dfrac{\sqrt{5}}{3}$

Ans. (B)

10. Approximate $\log_8 13$

    (A) 1.23
    (B) 1.21
    (C) 1.18
    (D) 1.15
    (E) 1.02

$\log_A B = \dfrac{\log B}{\log A}$

$\therefore \log_8 13 = \dfrac{\log 13}{\log 8} = 1.23$

Ans. (A)

11. The circumference of circle $x^2 + y^2 - 6x - 4y - 36 = 0$ is

    (A) 192
    (B) 38
    (C) 125
    (D) 54
    (E) 44

$x^2 - 6x + y^2 - 4y = 36 \rightarrow$ making a complete square form,
$(x - 3)^2 + (y - 2)^2 = 36 + 9 + 4 = 49 = 7^2$
$\therefore$ center $c = (3, 2)$, radius $r = 7$
$\therefore$ circumference $c = 2\pi r = 2\pi(7) = 14\pi = 43.98$

Ans. (E)

12. If $f(x) = 4$ for all real numbers x, then $f(x - 2) =$

    (A) 0
    (B) 2
    (C) 4
    (D) x
    (E) The value cannot be determined.

Notice that $y = f(x) = 4$ is just a horizontal line across all the values of x!!
Therefore, $f(x - 2) = 4$

Ans. (C)

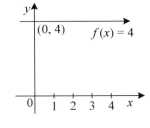

13. In right triangle ABC, $AB = 13$, $BC = 12$, $AC = 5$. The tangent of $\angle A$ is

(A) $\dfrac{12}{13}$ (B) $\dfrac{13}{12}$ (C) $\dfrac{12}{5}$ (D) $\dfrac{5}{12}$ (E) $\dfrac{5}{13}$

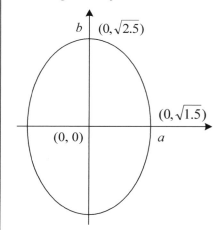

Referring to the figure, we get $\tan \theta = \dfrac{12}{5}$.

Ans. (C)

14. The sum of the roots of the equation
$(x + \sqrt{2})^2(2x - \sqrt{3})(x - \sqrt{7}) = 0$ is

(A) 3.3   (B) 0.7   (C) 2.1   (D) 6.3   (E) 2.5

The zeros of this equation are $x = -\sqrt{2},\ -\sqrt{2},\ \dfrac{\sqrt{3}}{2},\ \sqrt{7}$.

Therefore, the sum of all zeros is 0.68, or 0.7.

Ans. (B)

15. What is the length of the major axis of the ellipse whose equation is $10x^2 + 6y^2 = 15$?

(A) 1.26
(B) 2.50
(C) 3.16
(D) 4.47
(E) 5.00

$10x^2 + 6y^2 = 15 \rightarrow \dfrac{10x^2}{15} + \dfrac{6y^2}{15} = 1 \rightarrow$

$\dfrac{x^2}{\frac{15}{10}} + \dfrac{y^2}{\frac{15}{6}} = 1 \rightarrow \dfrac{x^2}{\left(\sqrt{1.5}\right)^2} + \dfrac{y^2}{\left(\sqrt{2.5}\right)^2} = 1$

$\therefore a = \sqrt{1.5},\ b = \sqrt{2.5}$

$\therefore$ The length of major axis $= 2b = 2\sqrt{2.5} = 3.16$

[ellipse figure with points b, $(0, \sqrt{2.5})$, $(0, \sqrt{1.5})$, $(0,0)$, a]

Ans. (C)

16. If $f(g(x)) = \dfrac{\frac{3}{2}\sqrt{x^2-1}-1}{2\sqrt{x^2-1}+1}$ and $f(x) = \dfrac{\frac{3}{2}x-1}{2x+1}$, then $g(x) =$

(A) $\sqrt{x}$
(B) $\sqrt{x^2 - 1}$
(C) x
(D) x^2
(E) $x^2 + 1$

$f(g(x)) = \dfrac{\frac{3}{2}\sqrt{x^2-1}-1}{2\sqrt{x^2-1}+1} \rightarrow \dfrac{\frac{3}{2}(2^3\sqrt{x^2-1}-1)}{2(2^3\sqrt{x^2-1}-1)}$

Now, by letting $u = \sqrt{x^2 - 1}$,

we get $f(u) = \dfrac{\frac{3}{2}u-1}{2u+1}$, or $f(x) = \dfrac{\frac{3}{2}x-1}{2x+1}$

$\therefore u = g(x) = \sqrt{x^2 - 1}$

Ans. (B)

17. If A is the degree measure of an acute angle and $\sin A = 0.25$, then $\sin A \cdot \cos (90° - A) =$

   (A) 0.02
   (B) 0.04
   (C) 0.05
   (D) 0.06
   (E) 0.08

Recall the property of Cofunction. $\sin (90° - A) = \cos A$
(eg) $\sin 30° = \cos 60°$, $\tan 25° = \cot 65°$, $\sec 15° = \csc 75°$, ...
Therefore, $\sin A \cdot \cos (90° - A)$
$= (\sin A)(\sin A) = (0.25)(0.25) = 0.0625$

Ans. (D)

18. The plane $x + 2y + 3z = 5$ intersects the x-axis at $(a,0,0)$, the y-axis at $(0,b,0)$, and the z-axis at $(0,0,c)$. What is the volume of the pyramid in the first octant with vertices, a, b, c and the origin?

   (A) 5
   (B) $\dfrac{125}{36}$
   (C) $\dfrac{65}{12}$
   (D) 1
   (E) 9

$x + 2y + 3z = 5$
Now, plugging in $x = a$, $y = 0$, $z = 0$, we get $a + 0 + 0 = 5$.
$\therefore a = 5$
For $(0, b, 0)$, $0 + 2b + 0 = 5$.
$\therefore b = \dfrac{5}{2}$
Also, for $(0, 0, c)$, $0 + 0 + 3c = 5$.
$\therefore c = \dfrac{5}{3}$

Thus, the base area in the xy-plane is $BA = \dfrac{1}{2} (5) \dfrac{5}{2} = \dfrac{25}{4}$.
Therefore, the volume,
$$V = \frac{1}{3} (BA)(H) = \frac{1}{3} \left(\frac{25}{4} \right) \left(\frac{5}{3} \right) = \frac{125}{36}$$

Ans. (B)

19. If $f(x,y) = x^2 - 2y^2$ and $g(x) = 3^x$, the value of $g(f(2,1)) =$

   (A) 1   (B) 4   (C) 9   (D) -4   (E) 0

$f(2, 1) = 2^2 - 2(1)^2 = 2$
$\therefore g(f(2, 1)) = g(2) = 3^2 = 9$

Ans. (C)

20. The units digit of 2567^{93} is

   (A) 1
   (B) 3
   (C) 7
   (D) 9
   (E) none of the above

$2567^1 = 2567 \rightarrow$ unit digit = 7
$2567^2 = \ldots 9 \rightarrow$ because $7 \times 7 = 49$ with unit digit 9
$2567^3 = \ldots 3 \rightarrow$ because $9 \times 7 = 63$ with unit digit 3
$2567^4 = \ldots 1 \rightarrow$ because $3 \times 7 = 21$ with unit digit 1
Now, we just found the cycle of the unit digit of being
$(7 \rightarrow 9 \rightarrow 3 \rightarrow 1 \rightarrow 7 \rightarrow 9 \rightarrow 3 \rightarrow 1 \rightarrow \ldots)$.
Since every four, it repeats, we divide the power of 93 by 4, and get the remainder R, which represent the number with unit digit that we are looking for.
That is, $93 \div 4 = 23$ with $R = 1$.
$\therefore$ The first one of the repetition of the unit digit, which is 7 in this case.

Ans. (C)

21. Solve for r: $5^{7-r} = 25^{r-1}$

   (A) 1
   (B) 2
   (C) 3
   (D) 4
   (E) 5

$5^{7-r} = 25^{r-1} \rightarrow 5^{7-r} = (5^2)^{r-1} = 5^{2r-2}$
$\therefore 7 - r = 2r - 2$
$\therefore 3r = 9$
$\therefore r = 3$

Ans. (C)

22. What is $\lim\limits_{x\to\infty} \dfrac{4x^2-\sqrt{x}+5}{3x^2+\sqrt{7}}$?

   (A) -1.12 (B) .91 (C) 1.11 (D) 1.33 (E) 1.50

Since the degree of the numerator and the degree of the denominator are the same with the degree 2, the limit will be the ratio of the coefficients of the highest degree.

$\therefore \dfrac{4}{3}$

Ans. (D)

23. A rod, pivoted at one end, rotates through $\dfrac{\pi}{4}$ radians.
   If the rod is 16 inches long, how many inches does the free end travel?

   (A) π
   (B) 2π
   (C) 3π
   (D) 4π
   (E) $\dfrac{3\pi}{2}$

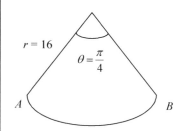

$\widehat{AB} = r\cdot\theta = 16\dfrac{\pi}{4} = 4\pi$

Ans. (D)

24. The graph of $y = 2x - |x|$ is equivalent to the graph of

   (A) $y = x$ for $x \le 0$
   (B) $y = x$ for $x \ge 0$
   (C) $y = 2x$ for $0 \le x \le 1$
   (D) $y = 2x$ for $x \le 0$
   (E) $y = 2x$ for $x \ge 0$

When $x \ge 0$, $y = 2x - x = x$
When $x < 0$, $y = 2x - (-x) = 3x$

Ans. (B)

25. If the radius of a sphere is doubled, the percent increase in surface area is

   (A) 100
   (B) 300
   (C) 400
   (D) 700
   (E) 800

$V = \dfrac{4}{3}\pi r^3$, $SA = 4\pi r^2$. Now, r is doubled. $\therefore 2r$

$\therefore V' = \dfrac{4}{3}\pi(2r)^3$, $SA' = 4\pi(2r)^2$
Since $SA' = 16\pi r^2$ vs. $SA = 4\pi r^2$, it is 4:1 ratio.
$\therefore$ the percent difference is 300%.

Ans. (B)

26. Find the value, in simplest form, of $\dfrac{(2^{2n})(2^{n+1})-(2^n)^3}{2(2^{3n})}$

   (A) $\dfrac{1}{2}$ (B) $\dfrac{1}{4}$ (C) $\dfrac{3}{4}$ (D) $\dfrac{5}{8}$ (E) $\dfrac{7}{8}$

$\dfrac{(2^{2n})(2^{n+1})-(2^n)^3}{2(2^{3n})} = \dfrac{2^{2n+n+1}-2^{3n}}{2^{3n+1}} =$

$\dfrac{2^{3n+1}-2^{3n}}{2^{3n+1}} = \dfrac{2^{3n}(2-1)}{2^{3n}\cdot 2} = \dfrac{1}{2}$

Ans. (A)

27. If $\log_9 p = 3$ and $\log_3 q = 4$, p expressed in terms of q is

   (A) $p = q^2$ (B) $p = q^8$ (C) $p = q^{\frac{3}{2}}$
   (D) $p = q^{\frac{5}{2}}$ (E) $p = q^3$

$\log_9 p = 3 \longrightarrow p = 9^3 = (3^2)^3 = 3^6$
$\log_3 q = 4 \longrightarrow q = 3^4$
$\therefore p = 3^6 = (3^4)^{\frac{3}{2}} = q^{\frac{3}{2}}$

Ans. (C)

28. If $a = 3\sec\theta$, and $b = 3\tan\theta$, then $\sqrt{a^2 - b^2}$

(A) 1
(B) 3
(C) 6
(D) $3\cos\theta\cdot\sin\theta$
(E) $3(\cos\theta + \sin\theta)$

Recall $1 + \tan^2\theta = \sec^2\theta$
$\therefore \sec^2\theta - \tan^2\theta = 1$
Since $\sqrt{a^2 - b^2} = \sqrt{9\sec^2\theta - 9\tan^2\theta}$
$= 3\sqrt{\sec^2\theta - \tan^2\theta} = 3\sqrt{1} = 3.$

Ans. (B)

29. The value of $\dfrac{525!}{521!3!}$ is

(A) greater than 10^{100}
(B) between 10^{10} and 10^{100}
(C) between 10^5 and 10^{10}
(D) between 10 and 10^5
(E) less than 10

$\dfrac{525!}{521!3!} = \dfrac{(525)(524)(523)(522)(521)(520)\,...3\cdot2\cdot1}{((521)(520)\,...3\cdot2\cdot1)\cdot(3\cdot2\cdot1)}$

$= \dfrac{(525)(524)(523)(522)}{(3\cdot2\cdot1)} = 1.2517 \times 10^{10}$

Ans. (B)

30. If $f(x) = ax + b$, which of the following make(s) $f(x) = f^{-1}(x)$?

I. $a = -1$, $b =$ any real number
II. $a = 1$, $b = 0$
III. $a =$ any real number, $b = 0$

(A) only I
(B) only II
(C) only III
(D) only I and II
(E) only I and III

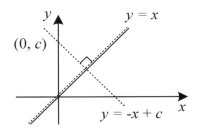

For $f(x)$ function to be the same as its inverse $f^{-1}(x)$, it must be on the same line as the symmetry line $y = x$, or perpendicular to $y = x$ line.
Therefore, $a = 1$, $b = 0$, or $a = -1$, $c =$ any #.

Ans. (D)

31. If the roots of the equation $ax^2 + bx + c = 0$ are α and β, the value of $\dfrac{(\alpha+\beta)^2}{\alpha\beta}$ in terms of a, b and c is

(A) $\dfrac{b^2}{a^2c^2}$ (B) $\dfrac{a^2c^2}{b}$ (C) $\dfrac{b^2}{ac}$ (D) $\dfrac{-b}{a^2c^2}$ (E) $\dfrac{-b^2}{ac}$

Recall from the properties of the quadratic function, we have sum of $\alpha + \beta = -\dfrac{b}{a}$, $\alpha\beta = \dfrac{c}{a}$.

$\therefore \dfrac{(\alpha+\beta)^2}{\alpha\beta} = \dfrac{\left(-\dfrac{b}{a}\right)^2}{\dfrac{c}{a}} = \left(-\dfrac{b}{a}\right)^2 \cdot \dfrac{a}{c} = \dfrac{b^2}{a^2}\cdot\dfrac{a}{c} = \dfrac{b^2}{ac}$

Ans. (C)

32. The range of the function $y = -x^{-\frac{2}{3}}$ is

(A) $y < 0$
(B) $y > 0$
(C) $y \geq 0$
(D) $y \leq 0$
(E) all real numbers

Using a graphing utility, we get $y < 0$.

Ans. (A)

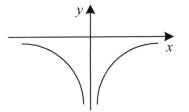

125

33. If P represents the set of rhombi and Q the set of squares, then the set $P \cap Q$ represents the set of

(A) squares
(B) trapezoids
(C) parallelograms
(D) quadrilaterals
(E) rectangles

The set of squares are included in the set of Rhombi, $\square \in \lozenge$. Therefore, the intersection of square and Rhombi is the set of squares.

Ans. (A)

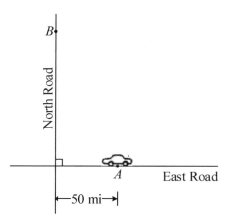

34. The figure above shows a car traveling from point A to point B. The straight-line distance between points A and B is 130 miles. If the car travels at an average speed of 65 miles per hour along East and North Roads, how long will it take the car to get to point B?

(A) 51 minutes
(B) 2 hour and 7 minutes
(C) 2 hour and 37 minutes
(D) 2 hour and 33 minutes
(E) 2 hour and 46 minutes

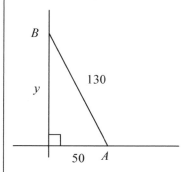

$\therefore y = \sqrt{130^2 - 50^2} = 120$
$\therefore$ Total distance $D = 50 + 120 = 170$
Speed, $S = 65$
$\therefore$ Time $T = \dfrac{D}{S} = \dfrac{170}{65} = 2.615$
or 2 hours and $(0.615) \times 60$ min = 2 hrs 37 min.

Ans. (C)

35. If y varies jointly as x, w, and the cube of z, what is the effect on w when x, y, and z are doubled?

(A) w is doubled
(B) w is multiplied by 4
(C) w is multiplied by 8
(D) w is divided by 8
(E) w is divided by 4

$y = k(x)(w)(z)^3$
Now, x, y, and z are doubled.
$\therefore 2y = k(2x)(?w)(2z)^3 \to y = k(x)(?w)(2z)^3 = k(x)(?w)(8z)^3$
This equation shows that the effect on w should be $\dfrac{w}{8}$,

or divided by 8, to get the same equation as the original equation.

Ans. (D)

36. $\sum_{k=1}^{4} 2\left(\dfrac{2}{3}\right)^k =$

(A) $3\dfrac{17}{81}$ (B) $4\dfrac{20}{27}$ (C) $5\dfrac{20}{27}$ (D) $4\dfrac{1}{3}$ (E) $5\dfrac{7}{27}$

$\sum_{k=1}^{4} 2\left(\dfrac{2}{3}\right)^k = 2\sum_{k=1}^{4} \left(\dfrac{2}{3}\right)^k = 2[\dfrac{2}{3} + \left(\dfrac{2}{3}\right)^2 + \left(\dfrac{2}{3}\right)^3 + \left(\dfrac{2}{3}\right)^4] = 3\dfrac{17}{81}$,

or using $s_4 = 2[a_1\left(\dfrac{1-r^4}{1-r}\right)] = 2[\dfrac{2}{3}(\dfrac{1-\left(\dfrac{2}{3}\right)^4}{1-\dfrac{2}{3}})] = 3\dfrac{17}{81}$.

Ans. (A)

37. $F(x) = \dfrac{5x^2 - 5}{x-1}$, when $x \neq 1$
    $= k$, when $x = 1$

    For what value(s) of k is F a continuous function?

    (A) 1
    (B) 2
    (C) 3
    (D) 10
    (E) no value of k

To be continuous, $f(1) = k$, but $f(x) = 5\left(\dfrac{x^2-1}{x-1}\right)$

$= \dfrac{5(x+1)(x-1)}{(x-1)} = 5(x+1)$,

where we had a removable discontinuity at $x = 1$.
Now, $f(1) = 5(1+1) = 10$
$\therefore k = 10$

Ans. (D)

38. What is the smallest positive x-intercept of the graph of $y = \dfrac{1}{2}\sin\left(4x + \dfrac{4\pi}{3}\right)$?

    (A) 0
    (B) 0.52
    (C) 1.05
    (D) 2.09
    (E) 1.31

Using a graphing utility,
$y_1 = \dfrac{1}{2}\sin\left(4x + \dfrac{4\pi}{3}\right)$ intercepts at $x = 0.52$.

Ans. (B)

39. How many integers greater than 2000 can be formed from the digits 1, 2, 3, and 4, if no digit is repeated in any number?

    (A) 9
    (B) 18
    (C) 27
    (D) 36
    (E) 72

2	3#'s	2#'s	1#'s	$= 3 \times 2 \times 1 = 6$
3	3#'s	2#'s	1#'s	$= 3 \times 2 \times 1 = 6$
4	3#'s	2#'s	1#'s	$= 3 \times 2 \times 1 = 6$

$\therefore$ Total of 18 possible combination

Ans. (B)

40. If $x(t) = 5\sin t$, $y(t) = 3 + 4\tan t$,
    what is the approximate value of x when $y = 6$?

    (A) 1
    (B) 3
    (C) 4
    (D) 5
    (E) 6

$y = 6$
$\therefore y(t) = 3 + 4\tan t = 6$
$\therefore 4\tan t = 3$
$\therefore \tan t = \dfrac{3}{4}$
$\therefore t = \tan^{-1}\dfrac{3}{4} = 36.9°$
$\therefore x = 5 \cdot \sin(36.9°) = 3$

Ans. (B)

41. If the roots of the equation $x^3 + px^2 + qx + r = 0$ are s_1, s_2 and s_3, then $s_1^2 + s_2^2 + s_3^2 =$

    (A) $p^2 + q^2$
    (B) $p^2 - 2q$
    (C) $p^2 - q^2$
    (D) p^2
    (E) q^2

Recall from the cubic equation, $ax^3 + bx^2 + cx + d = 0$,
the sum of 3 roots $x_1 + x_2 + x_3 = -\dfrac{b}{a}$,

$x_1x_2 + x_2x_3 + x_3x_1 = \dfrac{c}{a}$, $x_1x_2x_3 = -\dfrac{d}{a}$

In this problem, $a = 1$, $b = p$, $c = q$, $d = r$.
$\therefore s_1 + s_2 + s_3 = -p$, $s_1s_2 + s_2s_3 + s_3s_1 = q$
$\therefore s_1^2 + s_2^2 + s_3^2 = (s_1 + s_2 + s_3)^2 - 2(s_1s_2 + s_2s_3 + s_3s_1) =$
$(-p)^2 - 2q = p^2 - 2q$.

Ans. (B)

42. Find the area of the regular hexagon inscribed in a circle of radius 6.

(A) 36
(B) $36\sqrt{3}$
(C) 54
(D) $54\sqrt{3}$
(E) $128\sqrt{3}$

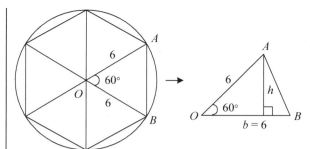

$h = 6\sin 60° = 3\sqrt{3}$

Area of $\triangle OAB = \dfrac{1}{2} \, b \cdot h = \dfrac{1}{2}(6)(3\sqrt{3}) = 9\sqrt{3}$

Since there are $6\triangle$'s, Area $= 6 \times 9\sqrt{3} = 54\sqrt{3}$

Ans. (D)

43. What is the period of $y = 5 - \cos 4x$?

(A) $\dfrac{\pi}{4}$

(B) $\dfrac{\pi}{2}$

(C) π

(D) $\dfrac{3\pi}{2}$

(E) 2π

Using a graphing calculator,

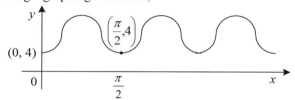

Refer to the above figure, $\cos 4x$ has its period, $p = \dfrac{2\pi}{4} = \dfrac{\pi}{2}$.

Ans. (B)

44. For what value of x is the function $\cos^2 x - \sin^2 x$ a minimum?

(A) 2π

(B) $\dfrac{3\pi}{4}$

(C) $\dfrac{\pi}{2}$

(D) $\dfrac{2\pi}{3}$

(E) $\dfrac{\pi}{4}$

Using the Double Angle property of Trig functions, we get $\cos 2\theta = \cos^2\theta - \sin^2\theta$. Referring to the following figure, using a graphing calculator; $p = \dfrac{2\pi}{2} = \pi$

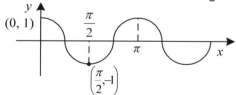

min. at $x = \dfrac{\pi}{2}$

Ans. (C)

45. What is the equation of the perpendicular bisector of the line segment whose end points are (-3,5) and (5,-1)?

(A) $4x - 3y + 2 = 0$
(B) $4x + 3y - 13 = 0$
(C) $x - 3y + 10 = 0$
(D) $x + 3y - 8 = 0$
(E) $4x + 3y - 1 = 3$

The definition of Perpendicular Bisector is "the set of all points that are equidistant from the two point of the line segment".

∴ The given equation must pass through its mid-point $M = (\dfrac{5-3}{2}, \dfrac{-1+5}{2}) = (1, 2)$.

The answer choice (A) $4x - 3y + 2 = 0$ is the only choice that satisfy $(1, 2)$.

Ans. (A)

46. If $3(x + 1)^2 + 10(y - 2)^2 = 48$ is graphed the sum of the distances from any point on the curve to the two foci is

   (A) 4
   (B) 8
   (C) 12
   (D) 16
   (E) 32

$3(x + 1)^2 + 10(y - 2)^2 = 48 \rightarrow \dfrac{(x+1)^2}{16} + \dfrac{(y-2)^2}{4.8} = 1$

$\therefore a = 4, b = \sqrt{4.8}$

$\therefore$ The length of major axis $= 2a = 8$, which is equal to the sum of the distances between two fixed points to a point on the curve.

Ans. (B)

47. Which of the following is the equation of the circle that has its center at $(1, 1)$ and is tangent to the line with equation $x + 3y = -6$?

   (A) $(x - 1)^2 + (y - 1)^2 = 2$
   (B) $(x - 1)^2 + (y - 1)^2 = 4$
   (C) $(x - 1)^2 + (y - 1)^2 = 3$
   (D) $(x - 1)^2 + (y - 1)^2 = 5$
   (E) $(x - 1)^2 + (y - 1)^2 = 10$

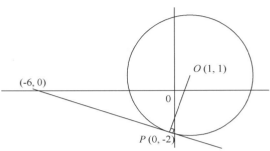

$\overline{OP} = \dfrac{|Ax_1 + By_1 + C|}{\sqrt{A^2 + B^2}}$, where $\overline{OP}$ is a radius for this circle.

Since $x + 3y + 6 = 0$, we get $A = 1, B = 3, C = 6$, and $x_1 = 1, y_1 = 1$.

$\therefore \overline{OP} = r = \dfrac{|1 + 3 + 6|}{\sqrt{10}} = \sqrt{10}$

$\therefore (x - 1)^2 + (y - 1)^2 = (\sqrt{10})^2 = 10$

Ans. (E)

48. The polar coordinates of a point P are $(2, \dfrac{4\pi}{3})$.

   The Cartesian (rectangular) coordinates of P are

   (A) $(-1, -\sqrt{3})$
   (B) $(-1, \sqrt{3})$
   (C) $(-\sqrt{3}, -1)$
   (D) $(-\sqrt{3}, 1)$
   (E) none of the above

Given $(r, \theta) = (2, \dfrac{4\pi}{3})$,

we have $x = r \cos \theta = 2 \cdot \cos \dfrac{4\pi}{3} = -1$,

and $y = r \sin \theta = 2 \cdot \sin \dfrac{4\pi}{3} = -\sqrt{3}$

$\therefore (x, y) = (-1, -\sqrt{3})$

Ans. (A)

49. If the magnitudes of vectors **u** and **v** are 4 and 8, respectively, then the magnitude of vector $(\mathbf{v} - \mathbf{u})$ could NOT be

   (A) 5   (B) 7   (C) 10   (D) 12   (E) 3

$|u| = 4, |v| = 8$.

Since $|v - u| \geq |v| - |u|$,

we get $|v - u| \geq 8 - 4$, or $|v - u| \geq 4$.

Ans. (E)

50. Given the statement "All boys play baseball," which of the following negates this statement?

   (A) All boys play tennis.
   (B) Some girls play tennis.
   (C) All boys do not play tennis.
   (D) At least one boy doesn't play baseball.
   (E) All girls do not play tennis.

The negation of "All" becomes "At least one".

Ans. (D)

Model Test No. 04

---

50 Questions / 60 Minutes

Directions: For each question, determine which of the answer choices is correct and fill in the oval on the answer sheet that corresponds to your choice.

Notes:

1. You will need to use a scientific or graphing calculator to answer some of the questions.
2. Be sure your calculator is in degree mode.
3. Each figure on this test is drawn as accurately as possible unless it is specifically indicated that the figure has not been drawn to scale.
4. The domain of any function f is the set of all real numbers x for which $f(x)$ is also a real number, unless the question indicates that the domain has been restricted in some way.
5. The box below contains five formulas that you may need to answer one or more of the questions.

REFERENCE INFORMATION

THE FOLLOWING INFORMATION IS FOR YOUR REFERENCE IN ANSWERING SOME OF THE QUESTIONS IN THIS TEST.

Volume of a right circular cone with radius r and height h: $V = \frac{1}{3}\pi r^2 h$

Lateral Area of a right circular cone with circumference of the base c and slant height l: $S = \frac{1}{2}cl$

Volume of a sphere with radius r: $V = \frac{4}{3}\pi r^3$

Surface Area of a sphere with radius r: $S = 4\pi r^2$

Volume of a pyramid with base area B and height h: $V = \frac{1}{3}Bh$

Answer Sheet
Model Test No. 04

1 Ⓐ Ⓑ Ⓒ Ⓓ Ⓔ	14 Ⓐ Ⓑ Ⓒ Ⓓ Ⓔ	27 Ⓐ Ⓑ Ⓒ Ⓓ Ⓔ	40 Ⓐ Ⓑ Ⓒ Ⓓ Ⓔ
2 Ⓐ Ⓑ Ⓒ Ⓓ Ⓔ	15 Ⓐ Ⓑ Ⓒ Ⓓ Ⓔ	28 Ⓐ Ⓑ Ⓒ Ⓓ Ⓔ	41 Ⓐ Ⓑ Ⓒ Ⓓ Ⓔ
3 Ⓐ Ⓑ Ⓒ Ⓓ Ⓔ	16 Ⓐ Ⓑ Ⓒ Ⓓ Ⓔ	29 Ⓐ Ⓑ Ⓒ Ⓓ Ⓔ	42 Ⓐ Ⓑ Ⓒ Ⓓ Ⓔ
4 Ⓐ Ⓑ Ⓒ Ⓓ Ⓔ	17 Ⓐ Ⓑ Ⓒ Ⓓ Ⓔ	30 Ⓐ Ⓑ Ⓒ Ⓓ Ⓔ	43 Ⓐ Ⓑ Ⓒ Ⓓ Ⓔ
5 Ⓐ Ⓑ Ⓒ Ⓓ Ⓔ	18 Ⓐ Ⓑ Ⓒ Ⓓ Ⓔ	31 Ⓐ Ⓑ Ⓒ Ⓓ Ⓔ	44 Ⓐ Ⓑ Ⓒ Ⓓ Ⓔ
6 Ⓐ Ⓑ Ⓒ Ⓓ Ⓔ	19 Ⓐ Ⓑ Ⓒ Ⓓ Ⓔ	32 Ⓐ Ⓑ Ⓒ Ⓓ Ⓔ	45 Ⓐ Ⓑ Ⓒ Ⓓ Ⓔ
7 Ⓐ Ⓑ Ⓒ Ⓓ Ⓔ	20 Ⓐ Ⓑ Ⓒ Ⓓ Ⓔ	33 Ⓐ Ⓑ Ⓒ Ⓓ Ⓔ	46 Ⓐ Ⓑ Ⓒ Ⓓ Ⓔ
8 Ⓐ Ⓑ Ⓒ Ⓓ Ⓔ	21 Ⓐ Ⓑ Ⓒ Ⓓ Ⓔ	34 Ⓐ Ⓑ Ⓒ Ⓓ Ⓔ	47 Ⓐ Ⓑ Ⓒ Ⓓ Ⓔ
9 Ⓐ Ⓑ Ⓒ Ⓓ Ⓔ	22 Ⓐ Ⓑ Ⓒ Ⓓ Ⓔ	35 Ⓐ Ⓑ Ⓒ Ⓓ Ⓔ	48 Ⓐ Ⓑ Ⓒ Ⓓ Ⓔ
10 Ⓐ Ⓑ Ⓒ Ⓓ Ⓔ	23 Ⓐ Ⓑ Ⓒ Ⓓ Ⓔ	36 Ⓐ Ⓑ Ⓒ Ⓓ Ⓔ	49 Ⓐ Ⓑ Ⓒ Ⓓ Ⓔ
11 Ⓐ Ⓑ Ⓒ Ⓓ Ⓔ	24 Ⓐ Ⓑ Ⓒ Ⓓ Ⓔ	37 Ⓐ Ⓑ Ⓒ Ⓓ Ⓔ	50 Ⓐ Ⓑ Ⓒ Ⓓ Ⓔ
12 Ⓐ Ⓑ Ⓒ Ⓓ Ⓔ	25 Ⓐ Ⓑ Ⓒ Ⓓ Ⓔ	38 Ⓐ Ⓑ Ⓒ Ⓓ Ⓔ	
13 Ⓐ Ⓑ Ⓒ Ⓓ Ⓔ	26 Ⓐ Ⓑ Ⓒ Ⓓ Ⓔ	39 Ⓐ Ⓑ Ⓒ Ⓓ Ⓔ	

1. If f is a linear function and $f(3) = 7, f(5) = 1,$ and $f(x) = 11.2$, what is the value of x?

    (A) 3.2
    (B) -1.9
    (C) 2.9
    (D) 1.6
    (E) -3.1

2. If $f(x) = \sqrt{x^2 - 1}$ and $g(x) = x + \dfrac{1}{x}$, then $g(f(3)) =$

    (A) 0.2
    (B) 1.2
    (C) 1.6
    (D) 3.2
    (E) 5.5

3. $\log(x^3 - y^3) =$

    (A) $\log x^3 - \log y^3$

    (B) $\log \dfrac{x^3}{y^3}$

    (C) $\log \dfrac{x+y}{x-y}$

    (D) $3\log x - 3\log y$
    (E) $\log (x - y) + \log (x^2 + xy + y^2)$

4. If $11^3 = 3^x$, what is the approximate value of x?

    (A) .95
    (B) 3.55
    (C) 5.95
    (D) 6.55
    (E) 7.95

5. The pendulum on a clock swings through an angle of $\dfrac{\pi}{2}$ radian, and the tip sweeps out an arc of 6 inches. How long is the pendulum?

    (A) 6 inches
    (B) 12 inches
    (C) 24 inches

    (D) $\dfrac{12}{\pi}$ inches

    (E) $\dfrac{24}{\pi}$ inches

6. $\left(-\dfrac{1}{8}\right)^{\frac{4}{3}} =$

(A) 0.06
(B) -0.25
(C) 6.35
(D) -0.06
(E) The value is not a real number

7. Which of the following is an (x, y) coordinate pair located on the ellipse $9x^2 + 25y^2 = 100$?

(A) $(1, 1.9)$
(B) $(1.5, 1.8)$
(C) $(2, 1.6)$
(D) $(2.5, 1.4)$
(D) $(3, 0.8)$

8. Where defined, $\sec \dfrac{\theta}{2} \cdot \cos \dfrac{\theta}{2} =$

(A) 1
(B) 0
(C) -1
(D) $2\csc 4\theta$
(E) $2\sec 4\theta$

9. Find the angle between $\dfrac{\pi}{2}$ and $\dfrac{3\pi}{2}$ that satisfies the equation: $2\cos^2 x - 7\sin x + 2 = 0$

(A) $\dfrac{2\pi}{3}$

(B) $\dfrac{5\pi}{6}$

(C) π

(D) $\dfrac{7\pi}{6}$

(E) $\dfrac{4\pi}{3}$

10. In figure 37, what is the approximate area of triangle *ABC*?

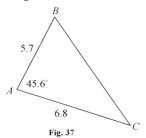

Fig. 37

   (A) 11.79
   (B) 13.85
   (C) 15.31
   (D) 17.10
   (E) 17.82

11. If $xy = -15$, $yz = -24$ and $xyz = -120$, then $y =$

   (A) -3
   (B) $\dfrac{3}{17}$
   (C) 1
   (D) 3
   (E) $\dfrac{17}{3}$

12. If $|2x - 4| < 10$, what are the possible values of x?

   (A) $0 < x < 5$
   (B) $0 < x < 2$
   (C) $-3 < x \leq 7$
   (D) $-3 \leq x < 7$
   (E) $-3 < x < 7$

13. If $0 \leq x \leq \dfrac{\pi}{2}$ and $\cos x = 5\sin x$, what is the value of x?

   (A) 0.197
   (B) 0.333
   (C) 0.613
   (D) 1.231
   (E) 1.373

14. If $f(x) = x^2 - 9$ is defined when $-3 \leq x \leq 3$, the maximum value of the graph of $|f(x)|$ is

   (A) -8
   (B) 0
   (C) 9
   (D) 4
   (E) 2

15. The graph of $y = |2x - 1| + 4$ consists of

    (A) one straight line
    (B) a pair of straight line rays
    (C) the sides of square
    (D) a circle
    (E) a parabola

16. In how many different ways can the seven letters in the word MAXIMUM be arranged, if all the letters are used each time?

    (A) 7
    (B) 42
    (C) 420
    (D) 840
    (E) 5040

17. If $13\sin x = 5$ and $\cos x < 0$, what is the approximate value of $\tan x$?

    (A) -0.42
    (B) -0.38
    (C) 0.38
    (D) 0.42
    (E) 0.92

18. When the graph of $y = \cos(3x)$ is drawn for all values of x between 0 and π, it crosses the x-axis

    (A) zero times
    (B) one times
    (C) two times
    (D) three times
    (E) six times

19. If $(1.73)^a = (1.39)^b$, what is the value of $\dfrac{a}{b}$?

    (A) -0.32
    (B) 0.32
    (C) 0.48
    (D) 0.60
    (E) 1.67

20. If the 10th term of an arithmetic sequence is 15 and the 20th term of the sequence is 100, what is the first term of the sequence?

    (A) -70
    (B) -61.5
    (C) 5
    (D) 61.5
    (E) 70

USE THIS SPACE FOR SCRATCH WORK

138

21. The set of points (x, y, z) such that $5x^2 + 5y^2 + 5z^2 = 1$ is

(A) empty
(B) a point
(C) a sphere
(D) a circle
(E) a plane

22. If a circle has a central angle of 45° that intercepts an arc of length 30 feet, the radius in feet is

(A) 63.7
(B) 38.2
(C) 44.1
(D) 75.0
(E) 28.6

23. In the figure, $\angle A = 100°$, $a = 7$ and $b = 3$. What is the value of $\angle C$?

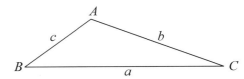

(A) 55°
(B) 25°
(C) 20°
(D) 15°
(E) 10°

24. If $f(x) = \sec x$ and $g(x) = 3x - 1$, which of the following is an even function (are even functions)?

I. $f(x) \cdot g(x)$
II. $f(g(x))$
III. $g(f(x))$

(A) only I
(B) only II
(C) only III
(D) only I and II
(E) only II and III

25. Which of the following statements is logically equivalent to: "If she exercises, she will pass the test."

(A) She passed the test; therefore, she exercised.
(B) She did not exercise;
     therefore, she will not pass the test.
(C) She did not pass the test;
     therefore she did not exercise.
(D) She will pass the test only if she exercises.
(E) None of the above.

26. Approximately, what is $\lim\limits_{x \to \sqrt{5}} \dfrac{2x^3 - x + 6}{x^2 + 3}$?

    (A) 1.42
    (B) 3.27
    (C) 5.38
    (D) 7.00
    (E) 8.67

27. If $\log_2 p = m$ and $\log_2 q = n$, then $pq =$

    (A) 2^{m+n}
    (B) 2^{mn}
    (C) 4^{mn}
    (D) 4^{m+n}
    (E) cannot be determined

28. If $f(x) = \sec x$, then

    (A) $f(x) = f(-x)$

    (B) $f\left(\dfrac{1}{x}\right) = -f(x)$

    (C) $f(-x) = -f(x)$

    (D) $f(x) = f\left(\dfrac{1}{x}\right)$

    (E) $f(x) = \dfrac{1}{f(x)}$

29. The graph of $|x| + |y| = 3$ consists of

    (A) one straight line
    (B) a pair of straight lines
    (C) the sides of a square
    (D) a circle
    (E) a point

30. In a circle an inscribed angle of 30° intercepts an arc of 10 inches. What is the area of the circle?

    (A) $\dfrac{3600}{\pi}$

    (B) $\dfrac{600}{\pi}$

    (C) $\dfrac{400}{\pi}$

    (D) 200π

    (E) not computable from given data

140

31. If P implies Q, an equivalent statement is

(A) Q implies P
(B) Q is a sufficient condition for P
(C) P is a sufficient condition for Q
(D) Not P implies Q
(E) Not P implies not Q

32. If $\log 2 = 0.30$ and $\log 11 = 1.04$,
what is the value of $\log_4 44$?

(A) 2.73
(B) 2.70
(C) 1.67
(D) 1.64
(E) 1.61

33. An isosceles triangle with base 10 and legs of 13 is
inscribed in a circle. Find the radius.

(A) 7.04
(B) 12
(C) 14.08
(D) 24
(E) cannot be determined

34. A regular octagon is formed by cutting off each corner
of a square. If the length of one side of the octagon is 5,
what is the area of the square?

(A) 450
(B) 366.42
(C) 225
(D) 145.71
(E) 72.86

35. If $2\sin^2 x - 3 = 3\cos x$ and $90° < x < 270°$,
the number of values that satisfy the equation is

(A) 0
(B) 1
(C) 2
(D) 3
(E) 4

36. If $A = Arc \cos \left(-\dfrac{3}{4} \right)$ and $A + B = 315°$, then $B =$

(A) 278.13°
(B) 176.41°
(C) -8.13°
(D) 171.87°
(E) 233.13°

USE THIS SPACE FOR SCRATCH WORK

37. If $f_{n+1} = f_{n-1} + 2 \cdot f_n$ for $n = 2, 3, 4, ...,$ and $f_1 = 1$ and $f_2 = 1$, then $f_5 =$

(A) 7
(B) 41
(C) 11
(D) 21
(E) 17

38. The fraction $\dfrac{i}{1+i}$ is equivalent to

(A) $1 - i$

(B) $\dfrac{1+i}{2}$

(C) $\dfrac{1-i}{2}$

(D) i

(E) $-i$

39. A root of $x^3 - 8 = 0$ lies in quadrant II. Write this root in polar form.

(A) $2(\cos 120° + i \sin 120°)$
(B) $2(\cos 144° + i \sin 144°)$
(C) $2(\cos 150° + i \sin 150°)$
(D) $4(\cos 144° + i \sin 144°)$
(E) $2(\cos 72° + i \sin 72°)$

40. The focus of a parabola is the point $(2,0)$ and its directrix is the line $x = -2$. Write an equation of the parabola.

(A) $y^2 = 8x$
(B) $x^2 = 8y$
(C) $x^2 = 4y$
(D) $y^2 = 4x$
(E) $x^2 = 2y$

41. If n is an integer, what is the remainder when $-2x^{2n+1} - 3x^{2n} + 3x^{2n-1} + 4$ is divided by $x + 1$?

(A) 0
(B) 2
(C) 4
(D) -8
(E) -13

42. Six men line up in a row. What is the probability that a certain two are next to each other?

(A) $\dfrac{1}{6}$

(B) $\dfrac{1}{4}$

(C) $\dfrac{1}{3}$

(D) $\dfrac{1}{2}$

(E) $\dfrac{2}{3}$

43. P varies directly as the cube of a and inversely as the square of b. If a is doubled and b is tripled, the value of P is

(A) multiplied by $\dfrac{3}{2}$

(B) multiplied by 6

(C) multiplied by $\dfrac{8}{9}$

(D) multiplied by 2

(E) divided by 2

44. Two roots of $3x^3 + 12x^2 + Kx - 36 = 0$ are equal numerically but opposite in sign. Find the value of K.

(A) -2
(B) +2
(C) -9
(D) +9

(E) $-\dfrac{9}{2}$

45. Two triangles having two equal sides and an equal angle between them

(A) are congruent
(B) are similar
(C) are equivalent
(D) have the same inscribed circle
(E) have the same circumscribed circle

46. If $x^2 + 4x + 3 < 0$ and $f(x) = x^2 - 4x - 3$, then

    (A) $0 < f(x) < 18$

    (B) $f(x) \geq \dfrac{3}{2}$

    (C) $f(x) \geq 12$

    (D) $f(x) \geq 0$

    (E) $2 < f(x) < 18$

47. The length of the vector that could correctly be used to represent in the complex plane the number $z = 1 + 3i$ is

    (A) 4
    (B) $\sqrt{10}$
    (C) 3
    (D) $2\sqrt{5}$
    (E) $\sqrt{11}$

48. If A and B are different points in a space, the set of all points in this space that are closer to A than to B is

    (A) the region of the space on one side of a plane
    (B) the interior of a sphere
    (C) a wedge-shaped region of the plane
    (D) the region of the space bounded by a cube
    (E) the interior of a circle

49. Which of the following lines are asymptotes of
    the graph of $y = \dfrac{x^2 + x + 1}{x}$?

    I. $x = 0$
    II. $y = 0$
    III. $y = x + 1$

    (A) I only
    (B) II only
    (C) I and II only
    (D) I and III only
    (E) I, II, and III

50. The radius of the base of a right circular cone is 5 and the radius of a parallel cross section is 3. If the distance between the base and the cross section is 6, what is the volume of the cone?

    (A) 57
    (B) 85
    (C) 157
    (D) 236
    (E) 393

Answer Key
Model Test No. 04

1	D	14	C	27	A	40	A
2	D	15	B	28	A	41	A
3	E	16	D	29	C	42	C
4	D	17	A	30	A	43	C
5	D	18	D	31	C	44	C
6	A	19	D	32	A	45	A
7	C	20	B	33	A	46	E
8	A	21	C	34	D	47	B
9	B	22	B	35	D	48	A
10	B	23	A	36	B	49	D
11	A	24	C	37	E	50	E
12	E	25	C	38	B		
13	A	26	B	39	A		

How to Score the SAT Subject Test in Mathematics Level 2

When you take an actual SAT Subject Test in Mathematics Level 2, your answer sheet will be "read" by a scanning machine that will record your responses to each question. Then a computer will compare your answers with the correct answers and produce your raw score. You get one point for each correct answer. For each wrong answer, you lose one-fourth of a point. Questions you omit (and any for which you mark more than one answer) are not counted. This raw score is converted to a scaled score that is reported to you and to the colleges you specify.

Finding Your Raw Test Score

STEP 1: Table A lists the correct answers for all the questions on the Subject Test in Mathematics Level 2 that is reproduced in this book. It also serves as a worksheet for you to calculate your raw score.

- Compare your answers with those given in the table.
- Put a check in the column marked "Right" if your answer is correct.
- Put a check in the column marked "Wrong" if your answer is incorrect.
- Leave both columns blank if you omitted the question.

STEP 2: Count the number of right answers.
Enter the total here: _____

STEP 3: Count the number of wrong answers.
Enter the total here: _____

STEP 4: Multiply the number of wrong answers by .250.
Enter the product here: _____

STEP 5: Subtract the result obtained in Step 4 from the total you obtained in Step 2.
Enter the result here: _____

STEP 6: Round the number obtained in Step 5 to the nearest whole number.
Enter the result here: _____

The number you obtained in Step 6 is your raw score.

Scaled Score Conversion Table
Subject Test in Mathematics Level 2

Raw Score	Scaled Score	Raw Score	Scaled Score	Raw Score	Scaled Score
50	800	28	630	6	470
49	800	27	630	5	460
48	800	26	620	4	450
47	800	25	610	3	440
46	800	24	600	2	430
45	800	23	600	1	420
44	800	22	590	0	410
43	790	21	580	-1	400
42	780	20	580	-2	390
41	770	19	570	-3	370
40	760	18	560	-4	360
39	750	17	560	-5	350
38	740	16	550	-6	340
37	730	15	540	-7	340
36	710	14	530	-8	330
35	700	13	530	-9	330
34	690	12	520	-10	320
33	680	11	510	-11	310
32	670	10	500	-12	300
31	660	9	490		
30	650	8	480		
29	640	7	480		

1. If f is a linear function and $f(3) = 7$, $f(5) = 1$, and $f(x) = 11.2$, what is the value of x?

   (A) 3.2
   (B) -1.9
   (C) 2.9
   (D) 1.6
   (E) -3.1

$f(x)$ is a liner function.
$\therefore f(x) = ax + b$
$\therefore f(3) = 3a + b = 7$, $f(5) = 5a + b = 1$
Therefore, $f(5) - f(3) = (5a + b) - (3a + b)$
$= 1 - 7 = (-)6$, or $2a = (-)6$
$\therefore a = (-)3$ $\therefore b = 16$
Since $f(x) = (-)3x + 16$, we get $(-)3x + 16 = 11.2$
$\therefore 3x = 4.8$
$\therefore x = 1.6$

Ans. (D)

2. If $f(x) = \sqrt{x^2 - 1}$ and $g(x) = x + \dfrac{1}{x}$, then $g(f(3)) =$

   (A) 0.2   (B) 1.2   (C) 1.6   (D) 3.2   (E) 5.5

$f(3) = \sqrt{3^2 - 1} = \sqrt{8}$
$\therefore g(f(3)) = g(\sqrt{8}) = \sqrt{8} + \dfrac{1}{\sqrt{8}} = 3.2$

Ans. (D)

3. $\log(x^3 - y^3) =$

   (A) $\log x^3 - \log y^3$ (B) $\log \dfrac{x^3}{y^3}$
   (C) $\log \dfrac{x+y}{x-y}$ (D) $3\log x - 3\log y$
   (E) $\log (x - y) + \log (x^2 + xy + y^2)$

$\log (x^3 - y^3) = \log (x - y)(x^2 + xy + y^2)$
Now, referring to the property of logarithmic function,
$\log A \cdot B = \log A + \log B$,
we get $\log (x - y)(x^2 + xy + y^2) = \log (x - y) + \log (x^2 + xy + y^2)$

Ans. (E)

4. If $11^3 = 3^x$, what is the approximate value of x?

   (A) .95
   (B) 3.55
   (C) 5.95
   (D) 6.55
   (E) 7.95

Since $11^3 = 3^x$ is an exponential equation, to bring the variable
x down, we need to apply the natural log on both sides.
$\therefore \ln 11^3 = \ln 3^x$
$\therefore 3 \cdot \ln 11 = x \cdot \ln 3$
$\therefore x = \dfrac{3 \ln 11}{\ln 3} = 6.55$

Ans. (D)

5. The pendulum on a clock swings through an angle of $\dfrac{\pi}{2}$ radian, and the tip sweeps out an arc of 6 inches. How long is the pendulum?

   (A) 6 inches
   (B) 12 inches
   (C) 24 inches
   (D) $\dfrac{12}{\pi}$ inches
   (E) $\dfrac{24}{\pi}$ inches

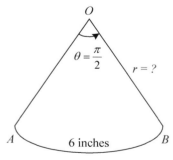

Referring to the figure next, we get $\overset{\frown}{AB} = 6 = r \cdot \dfrac{\pi}{2}$
(Note: $\overset{\frown}{AB} = r \cdot \theta$, where θ is in Radian mode)
$\therefore r = \dfrac{12}{\pi}$

Ans. (D)

148

6. $\left(-\dfrac{1}{8}\right)^{\frac{4}{3}} =$

   (A) 0.06 (B) -0.25 (C) 6.35 (D) -0.06
   (E) The value is not a real number

Using calculator, we get $-\dfrac{1}{8} \wedge \dfrac{4}{3} = 0.06$

Or, $\left(\sqrt[3]{\left(-\dfrac{1}{2}\right)^3}\right)^4 = \left(-\dfrac{1}{2}\right)^4 = \dfrac{1}{16} = 0.06$

Ans. (A)

7. Which of the following is an (x, y) coordinate pair located on the ellipse $9x^2 + 25y^2 = 100$?

   (A) (1, 1.9) (B) (1.5, 1.8) (C) (2, 1.6)
   (D) (2.5, 1.4) (D) (3, 0.8)

Since the point (x, y) has to be on the graph of $9x^2 + 25y^2 = 100$, we try answer choices of (A), (B), …
The only point that satisfy the equation is (C) (2, 1.6)

Ans. (C)

8. Where defined, $\sec\dfrac{\theta}{2} \cdot \cos\dfrac{\theta}{2} =$

   (A) 1 (B) 0 (C) -1
   (D) $2\csc 4\theta$ (E) $2\sec 4\theta$

$\sec\dfrac{\theta}{2} = \dfrac{1}{\cos\dfrac{\theta}{2}}$

$\therefore \sec\dfrac{\theta}{2} \cdot \cos\dfrac{\theta}{2} = \dfrac{\cos\dfrac{\theta}{2}}{\cos\dfrac{\theta}{2}} = 1$

Ans. (A)

9. Find the angle between $\dfrac{\pi}{2}$ and $\dfrac{3\pi}{2}$ that satisfies the equation: $2\cos^2 x - 7\sin x + 2 = 0$

   (A) $\dfrac{2\pi}{3}$

   (B) $\dfrac{5\pi}{6}$

   (C) π

   (D) $\dfrac{7\pi}{6}$

   (E) $\dfrac{4\pi}{3}$

Given $\dfrac{\pi}{2} < \theta < \dfrac{3\pi}{2}$, which is II - III quadrants,
$2\cos^2 x - 7\sin x + 2 = 2(1 - \sin^2 x) - 7\sin x + 2$
$= -2\sin^2 x - 7\sin x + 4 = 0$
$\therefore (-2\sin x + 1)(\sin x + 4) = 0$
$\therefore -2\sin x + 1 = 0$, or $\sin x = \dfrac{1}{2}$
$\therefore x = \sin^{-1}\dfrac{1}{2} = \dfrac{\pi}{6}$ or $\dfrac{5\pi}{6}$.
But we are in (II) or (III) quadrants, we chose $x = \dfrac{5\pi}{6}$.
Also, $\sin x + 4 = 0$, which is never possible, since $-1 \le \sin x \le 1$.
$\therefore$ The answers is $x = \dfrac{5\pi}{6}$.

Ans. (B)

10. In figure 37, what is the approximate area of triangle ABC?

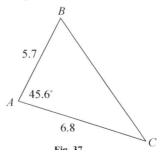

Fig. 37

(A) 11.79 (B) 13.85 (C) 15.31 (D) 17.10 (E) 17.82

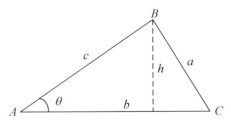

The area of $\triangle ABC$ is $\dfrac{1}{2}bh = \dfrac{1}{2}bc\sin\theta$, where $h = c \cdot \sin\theta$.
Since we have $b = 6.8$, $c = 5.7$ and $\theta = 45.6°$,
we have area $\triangle ABC = \dfrac{1}{2} \cdot (6.8)(5.7)\sin 45.6° = 13.85$.

Ans. (B)

11. If $xy = -15$, $yz = -24$ and $xyz = -120$, then $y =$

 (A) -3
 (B) $\dfrac{3}{17}$
 (C) 1
 (D) 3
 (E) $\dfrac{17}{3}$

$\dfrac{xyz}{xy} = z = \dfrac{(-)120}{(-)15} = 8$, $\dfrac{xyz}{yz} = x = \dfrac{(-)120}{(-)24} = 5$

$\therefore$ with $z = 8$, and $x = 5$, we get $xyz = (5)y(8) = (-)120$

$\therefore y = -3$

Or $(xy)(yz) = (-15)(-24) = xy^2z = 360$

$\therefore \dfrac{xy^2z}{xyz} = y = \dfrac{360}{(-)120} = -3$

$\therefore y = -3$

Ans. (A)

12. If $|2x - 4| < 10$, what are the possible values of x?

 (A) $0 < x < 5$
 (B) $0 < x < 2$
 (C) $-3 < x \le 7$
 (D) $-3 \le x < 7$
 (E) $-3 < x < 7$

$|2x - 4| < 10$

$\therefore -10 < 2x - 4 < 10$

$\therefore -6 < 2x < 14$

$\therefore -3 < x < 7$

Ans. (E)

13. If $0 \le x \le \dfrac{\pi}{2}$ and $\cos x = 5\sin x$, what is the value of x?

 (A) 0.197
 (B) 0.333
 (C) 0.613
 (D) 1.231
 (E) 1.373

$\cos x = 5\sin x$

$\therefore \dfrac{1}{5} = \dfrac{\sin x}{\cos x} = \tan x$

$\therefore x = \tan^{-1}\dfrac{1}{5} = 0.197$ radian

Ans. (A)

14. If $f(x) = x^2 - 9$ is defined when $-3 \le x \le 3$, the maximum value of the graph of $|f(x)|$ is

 (A) -8
 (B) 0
 (C) 9
 (D) 4
 (E) 2

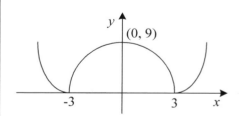

Using graphing utility, let $y_1 = |x^2 - 9|$.
Since our domain is $-3 \le x \le 3$, we get the max of $|f(x)| = 9$.

Ans. (C)

15. The graph of $y = |2x - 1| + 4$ consists of

 (A) one straight line
 (B) a pair of straight line rays
 (C) the sides of square
 (D) a circle
 (E) a parabola

Again, using graphing utility,
we get $y_1 = |2x - 1| + 4$ with the following graph.

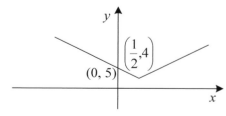

Ans. (B)

16. In how many different ways can the seven letters in the word MAXIMUM be arranged, if all the letters are used each time?

 (A) 7
 (B) 42
 (C) 420
 (D) 840
 (E) 5040

Refer to the previous note on "Probability and Combination", where COFFEE $= \dfrac{6!}{2!2!}$.

Since "MAXIMUM" has 7 letters with 3M's repeating, we get $\dfrac{7!}{3!} = 840$.

Ans. (D)

17. If $13\sin x = 5$ and $\cos x < 0$, what is the approximate value of $\tan x$?

 (A) -0.42
 (B) -0.38
 (C) 0.38
 (D) 0.42
 (E) 0.92

$13\sin x = 5$

$\therefore \sin x = \dfrac{5}{13}$, or $x = \sin^{-1}\dfrac{5}{13} = 22.6°$

But $\cos x < 0$

$\therefore \sin x = (+)$ and $\cos x = (-)$ implies that it is in the (II) quadrant.

$\therefore x = 180 - 22.6° = 157.4°$

$\therefore \tan x = \tan 157.4° = (-)0.42$

Ans. (A)

18. When the graph of $y = \cos(3x)$ is drawn for all values of x between 0 and π, it crosses the x-axis

 (A) zero times
 (B) one times
 (C) two times
 (D) three times
 (E) six times

Using graphing utility, let $y_1 = \cos(3x)$, and set our window, $0 < x < \pi$, then we find that it crosses x-axis 3 times, as shown in the following graph.

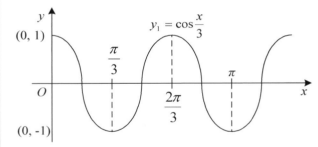

Ans. (D)

19. If $(1.73)^a = (1.39)^b$, what is the value of $\dfrac{a}{b}$?

 (A) -0.32
 (B) 0.32
 (C) 0.48
 (D) 0.60
 (E) 1.67

$(1.73)^a = (1.39)^b$

$\therefore \ln(1.73)^a = \ln(1.39)^b$, or $a\cdot\ln(1.73) = b\cdot\ln(1.39)$

$\therefore \dfrac{a}{b} = \dfrac{\ln 1.39}{\ln 1.73} = 0.60$

Ans. (D)

20. If the 10th term of an arithmetic sequence is 15 and the 20th term of the sequence is 100, what is the first term of the sequence?

 (A) -70
 (B) -61.5
 (C) 5
 (D) 61.5
 (E) 70

Referring to the previous note on "Sequence and Series", where $a_n = a_1 + (n - 1)d$,

$a_{10} = a_1 + 9d = 15$, $a_{20} = a_1 + 19d = 100$

$\therefore a_{20} - a_{10} = 10d = 85$

$\therefore d = 8.5$

$\therefore a_1 = 15 - 9d = 15 - 9(8.5) = (-)61.5$

Ans. (B)

21. The set of points (x, y, z) such that $5x^2 + 5y^2 + 5z^2 = 1$ is

    (A) empty
    (B) a point
    (C) a sphere
    (D) a circle
    (E) a plane

$5x^2 + 5y^2 + 5z^2 = 1 \rightarrow 5(x^2 + y^2 + z^2) = 1$, or $x^2 + y^2 + z^2 = \frac{1}{5}$,

which is the sphere of center $(0, 0, 0)$ and $r = \frac{1}{\sqrt{5}}$.

Ans. (C)

22. If a circle has a central angle of 45° that intercepts an arc of length 30 feet, the radius in feet is

    (A) 63.7
    (B) 38.2
    (C) 44.1
    (D) 75.0
    (E) 28.6

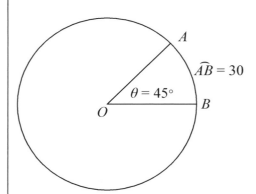

Using $\widehat{AB} = r \cdot \theta$, we get $30 = r \cdot \frac{\pi}{4}$, where $\theta = 45° = \frac{\pi}{4}$ radian.

$\therefore r = 30 \times \frac{4}{\pi} = \frac{120}{\pi} = 38.2$

Ans. (B)

23. In the figure, $\angle A = 100°$, $a = 7$ and $b = 3$. What is the value of $\angle C$?

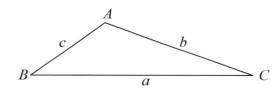

    (A) 55°
    (B) 25°
    (C) 20°
    (D) 15°
    (E) 10°

Using the law of sine, $\dfrac{\sin A}{a} = \dfrac{\sin B}{b} = \dfrac{\sin C}{c}$,

we get $\dfrac{\sin 100°}{7} = \dfrac{\sin B}{3}$.

$\therefore \sin B = 0.422$

$\therefore B = \sin^{-1}(0.422) = 25°$

$\therefore \angle C = 180° - 100° - 25° = 55°$

Ans. (A)

24. If $f(x) = \sec x$ and $g(x) = 3x - 1$, which of the following is an even function (are even functions)?

    I. $f(x) \cdot g(x)$
    II. $f(g(x))$
    III. $g(f(x))$

    (A) only I
    (B) only II
    (C) only III
    (D) only I and II
    (E) only II and III

Since even function is symmetric to y-axis, using graphing utility, let $f(x)g(x)$ as $y_1 = (\sec x)(3x - 1) = \dfrac{1}{\cos x} \cdot (3x - 1)$,

and $f(g(x))$ as, $y_2 = \sec(3x - 1) = \dfrac{1}{\cos(3x-1)}$.

Also, $g(f(x))$ as, $y_3 = 3(\sec x) - 1 = \dfrac{3}{\cos x} - 1$.

Then, we find that (III) $g(f(x)) = 3(\sec x) - 1$ is the only one that is symmetric to y-axis, thus it is even function.

Ans. (C)

152

25. Which of the following statements is logically equivalent to: "If she exercises, she will pass the test."

  (A) She passed the test; therefore, she exercised.
  (B) She did not exercise;
      therefore, she will not pass the test.
  (C) She did not pass the test;
      therefore she did not exercise.
  (D) She will pass the test only if she exercises.
  (E) None of the above.

In this problem, "logically equivalent" means to find "contra positive". Referring to our previous note on "logic", the contra positive of the statement $p \rightarrow q$ is $\sim q \rightarrow \sim p$, where "$\sim$" means negation.
∴ "If she exercise, she will pass the test." Becomes "If she does not pass the test, then she did not exercise."

Ans. (C)

26. Approximately, what is $\lim\limits_{x \to \sqrt{5}} \dfrac{2x^3 - x + 6}{x^2 + 3}$?

  (A) 1.42   (B) 3.27   (C) 5.38
  (D) 7.00   (E) 8.67

Simply by replacing $x = \sqrt{5}$, we get the result 3.27.

Ans. (B)

27. If $\log_2 p = m$ and $\log_2 q = n$, then $pq =$

  (A) 2^{m+n}
  (B) 2^{mn}
  (C) 4^{mn}
  (D) 4^{m+n}
  (E) cannot be determined

Referring to our previous note on "log" properties,
$\log_2 p = m \rightarrow p = 2^m$, $\log_2 q = n \rightarrow q = 2^n$
∴ $p \cdot q = 2^m \cdot 2^n = 2^{m+n}$

Ans. (A)

28. If $f(x) = \sec x$, then

  (A) $f(x) = f(-x)$

  (B) $f\left(\dfrac{1}{x}\right) = -f(x)$

  (C) $f(-x) = -f(x)$

  (D) $f(x) = f\left(\dfrac{1}{x}\right)$

  (E) $f(x) = \dfrac{1}{f(x)}$

$f(x) = \sec x = \dfrac{1}{\cos x}$, which is even function, where $f(-x) = f(x)$ is the definition of even function.

Ans. (A)

29. The graph of $|x| + |y| = 3$ consists of

  (A) one straight line
  (B) a pair of straight lines
  (C) the sides of a square
  (D) a circle
  (E) a point

The graph of $|x| + |y| = 3$ is as shown here. Therefore,

Ans. (C)

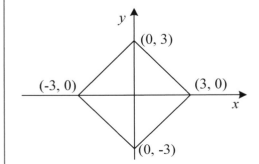

30. In a circle an inscribed angle of 30° intercepts an arc of 10 inches. What is the area of the circle?

(A) $\dfrac{3600}{\pi}$

(B) $\dfrac{600}{\pi}$

(C) $\dfrac{400}{\pi}$

(D) 200π

(E) not computable from given data

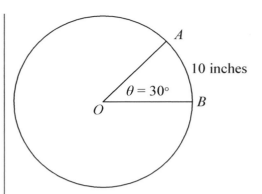

$\widehat{AB} = r \cdot \theta$

$\therefore 10 = r(30°) = r\left(\dfrac{\pi}{6}\right)$

$\therefore r = \dfrac{60}{\pi}$

Therefore, the area $= \pi r^2 = \pi\left(\dfrac{60}{\pi}\right)^2 = \dfrac{3600}{\pi}$.

Ans. (A)

31. If P implies Q, an equivalent statement is

(A) Q implies P
(B) Q is a sufficient condition for P
(C) P is a sufficient condition for Q
(D) Not P implies Q
(E) Not P implies not Q

In logic, P implies Q suggest had $P \in Q$, where P is the subset of the set Q, or as shown in the following figure, the smaller set P is a sufficient condition for the larger set Q, whereas the larger set Q is a necessary condition for the smaller set P.

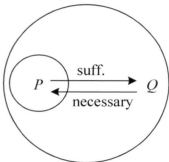

Ans. (C)

32. If $\log 2 = 0.30$ and $\log 11 = 1.04$, what is the value of $\log_4 44$?

(A) 2.73
(B) 2.70
(C) 1.67
(D) 1.64
(E) 1.61

$\log_4 44 = \dfrac{\log 44}{\log 4} = \dfrac{\log(2\times2\times11)}{\log(2\times2)}$

$= \dfrac{\log 2 + \log 2 + \log 11}{\log 2 + \log 2} = \dfrac{1.64}{0.60} = 2.73$

Ans. (A)

33. An isosceles triangle with base 10 and legs of 13 is inscribed in a circle. Find the radius.

(A) 7.04
(B) 12
(C) 14.08
(D) 24
(E) cannot be determined

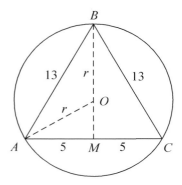

Given dimensions in above figure,
we get $\overline{BM} = \sqrt{13^2 - 5^2} = 12$.
Now, considering $\triangle ABM$, with $\overline{OA} = \overline{OB}$ = radius r,
we get $\overline{OM} = \overline{BM} - \overline{OB} = 12 - r$.

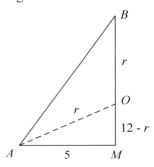

∴ $\triangle AOM$ gives, $r^2 = 5^2 + (12 - r)^2$
∴ $r^2 = 25 + 144 - 24r + r^2$, or $24r = 169$
∴ $r = \dfrac{169}{24} = 7.04$

Ans. (A)

34. A regular octagon is formed by cutting off each corner of a square. If the length of one side of the octagon is 5, what is the area of the square?

(A) 450
(B) 366.42
(C) 225
(D) 145.71
(E) 72.86

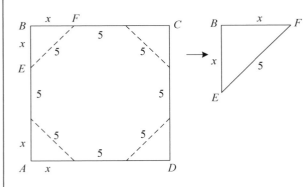

∴ $x^2 + x^2 = 2x^2 = 5^2$
∴ $x^2 = \dfrac{25}{2} = 12.5$
∴ $x = \sqrt{12.5}$
∴ $\overline{AB} = x + 5 + x = 2x + 5 = 2\sqrt{12.5} + 5 = 12.07$
∴ Area $\square ABCD = (12.07)^2 = 145.71$

Ans. (D)

155

35. If $2\sin^2 x - 3 = 3\cos x$ and $90° < x < 270°$, the number of values that satisfy the equation is

(A) 0
(B) 1
(C) 2
(D) 3
(E) 4

In this problem, let's use graphing utility!
Let $y_1 = 2\sin^2 x - 3 - 3\cos x$, and set the window as
$90° < x < 270°$, or $\dfrac{\pi}{2} < x < \dfrac{3\pi}{2}$.
We want to check now many x- intercepts are in this domain.
We get three intercepts.

Ans. (D)

Or, $y = 2\sin^2 x - 3\cos x - 3 = 2(1 - \cos^2 x) - 3\cos x - 3$
$= -2\cos^2 x - 3\cos x - 1 = 0$.
Or $(2\cos x + 1)(\cos x + 1) = 0$
$\therefore \cos x = -\dfrac{1}{2}$
$\therefore x = \dfrac{2\pi}{3}$ and $\dfrac{4\pi}{3}$ and also, $\cos x = -1$, which is at $x = \pi$.
$\therefore$ Altogether $x = \dfrac{2\pi}{3}, \pi, \dfrac{4\pi}{3}$

36. If $A = Arc \cos\left(-\dfrac{3}{4}\right)$ and $A + B = 315°$, then $B =$

(A) 278.13°
(B) 176.41°
(C) -8.13°
(D) 171.87°
(E) 233.13°

$A = \cos^{-1} -\dfrac{3}{4} = 138.6°$
$\therefore A + B = 138.6° + B = 315°$
$\therefore B = 315° - 138.6° = 176.40°$

(Please note that angle A also could be in the 3rd quadrant with $180° + (180° - 138.6°) = 221.4°$
$\therefore 221.4° + B = 315°$
$\therefore B = 93.6°$)

Ans. (B)

37. If $f_{n+1} = f_{n-1} + 2 \cdot f_n$ for $n = 2, 3, 4, ...,$ and $f_1 = 1$ and $f_2 = 1$, then $f_5 =$

(A) 7
(B) 41
(C) 11
(D) 21
(E) 17

Given $f_{n+1} = f_{n-1} + 2f_n$, $f_3 = f_1 + 2f_2 = 1 + 2(1) = 3$,
$f_4 = f_2 + 2f_3 = 1 + 2(3) = 1 + 2(3) = 7$
$f_5 = f_3 + 2f_4 = 3 + 2(7) = 17$

Ans. (E)

38. The fraction $\dfrac{i}{1+i}$ is equivalent to

(A) $1 - i$

(B) $\dfrac{1+i}{2}$

(C) $\dfrac{1-i}{2}$

(D) i

(E) $-i$

By multiplication of the conjugate of $(1 + i)$,
we get $\dfrac{i}{1+i} = \dfrac{i(1-i)}{(1+i)(1-i)} = \dfrac{(1-i^2)}{(1^2-i^2)} = \dfrac{i-(-1)}{1-(-1)} = \dfrac{1+i}{2}$.

Ans. (B)

39. A root of $x^3 - 8 = 0$ lies in quadrant II. Write this root in polar form.

   (A) $2(\cos 120° + i \sin 120°)$
   (B) $2(\cos 144° + i \sin 144°)$
   (C) $2(\cos 150° + i \sin 150°)$
   (D) $4(\cos 144° + i \sin 144°)$
   (E) $2(\cos 72° + i \sin 72°)$

Referring to our previous note on "Complex number" in polar

form, $z = a + bi = r(\cos \theta + i \sin \theta)$, where $r = \sqrt{a^2 + b^2}$, $\theta = \tan^{-1} \dfrac{b}{a}$, we first need to convert $x^3 = 8$ into $x^3 = 8 + 0i$,

where $a + bi = 8 + 0i$,

$\therefore a = 8, b = 0$.

The polar form becomes, $x^3 = 8 + 0i = r(\cos \theta + i \sin \theta)$,

where $r = \sqrt{8^2 + 0^2} = 8$, $\theta = \tan^{-1} \dfrac{0}{8} = 0°$ or $360°$

(Please note that $x^3 = 8 + 0i$ lies on the positive x-axis)

$\therefore x^3 = 8[\cos 360° + i \sin 360°]$.

Now, using De Moivre's Theorem,

$x = [8(\cos 360° + i \sin 360°)]^{\frac{1}{3}} = $

$8^{\frac{1}{3}} [\cos \dfrac{1}{3}(360°) + i \sin \dfrac{1}{3}(360°)] = $

$2(\cos 120° + i \sin 120°)$.

Ans. (A)

But processing with this mathematical approach takes too long!! Therefore, I strongly recommend to use calculator by checking answer choices (A), (B), …

The answer choice (A) gives, $[2(\cos 120° + i \sin 120°)]^3 = 8$.

40. The focus of a parabola is the point (2,0) and its directrix is the line $x = -2$. Write an equation of the parabola.

   (A) $y^2 = 8x$
   (B) $x^2 = 8y$
   (C) $x^2 = 4y$
   (D) $y^2 = 4x$
   (E) $x^2 = 2y$

Referring to our previous note on "Conic Sections", the equation of parabola is $4py = x^2$ or $4px = y^2$.

Since directrix line is $x = (-)2$, we use $4px = y^2$,

where the focus point is on x-axis with $F = (p, 0) = (2, 0)$.

$\therefore p = 2$

$\therefore 4(2)x = y^2$, which also have directrix $x = -p = -2$.

Ans. (A)

41. If n is an integer, what is the remainder when $-2x^{2n+1} - 3x^{2n} + 3x^{2n-1} + 4$ is divided by $x + 1$?

   (A) 0
   (B) 2
   (C) 4
   (D) -8
   (E) -13

Referring to our note on "remainder Theorem", where $p(a)$ = remainder, when $p(x)$ is divided by $(x - a)$.

$\therefore p(-1) = -2(-1)^{2n+1} - 3(-1)^{2n} + 3(-1)^{2n-1} + 4 = 2 - 3 - 3 + 4 = 0$,

which really means $(x + 1)$ is a factor of $p(x)$.

Ans. (A)

42. Six men line up in a row. What is the probability that a certain two are next to each other?

   (A) $\dfrac{1}{6}$ (B) $\dfrac{1}{4}$ (C) $\dfrac{1}{3}$ (D) $\dfrac{1}{2}$ (E) $\dfrac{2}{3}$

Referring to our note on "combination", we have A-B-C-D-E-F, where let's say (A-B) are always together, then (A-B)-C-D-E-F. Here, we consider (A-B) as one group, then the combination becomes 5!.

But (A-B) and (B-A) serves as 2!

$\therefore (5! \times 2!)$ out of total possible combination 6!

$\therefore p = \dfrac{5! \times 2!}{6!} = \dfrac{1}{3}$

Ans. (C)

43. P varies directly as the cube of a and inversely as the square of b. If a is doubled and b is tripled, the value of P is

(A) multiplied by $\dfrac{3}{2}$

(B) multiplied by 6

(C) multiplied by $\dfrac{8}{9}$

(D) multiplied by 2

(E) divided by 2

$P = k \cdot a^3 \cdot \dfrac{1}{b^2}$

$\therefore p' = k(2a)^3 \cdot \dfrac{1}{(3b)^2} = k(8a^3) \cdot \dfrac{1}{9b^2} = \dfrac{8}{9}\left(k \cdot a^3 \cdot \dfrac{1}{b^2}\right) = \dfrac{8}{9}P$

Ans. (C)

---

44. Two roots of $3x^3 + 12x^2 + Kx - 36 = 0$ are equal numerically but opposite in sign. Find the value of K.

(A) -2
(B) +2
(C) -9
(D) +9
(E) $-\dfrac{9}{2}$

Also, referring to our note on "sum and product of the roots", $ax^3 + bx^2 + cx + d = 0$,

then the sum $x_1 + x_2 + x_3 = -\dfrac{b}{a}$, $x_1 \cdot x_2 \cdot x_3 = -\dfrac{d}{a}$.

But $x_3 = -x_1$, $\therefore x_1 + x_2 + (-x_1) = -\dfrac{b}{a}$.

Here, $a = 3$, $b = 12$, $c = k$, $d = -36$,

$\therefore x_1 + x_2 + x_3 = x_2 = -\dfrac{12}{3} = -4$

Putting back $x = -4$ into $p(x)$,

we get $p(-4) = 3(-4)^3 + 12(-4)^2 - 4k - 36 = 0$.

$\therefore -192 + 192 - 4k - 36 = 0$

$\therefore k = -9$

Ans. (C)

---

45. Two triangles having two equal sides and an equal angle between them

(A) are congruent
(B) are similar
(C) are equivalent
(D) have the same inscribed circle
(E) have the same circumscribed circle

To be congruent triangles, we have
1) side-side-side
2) side-angle-side
3) angle-side-angle.

Ans. (A)

---

46. If $x^2 + 4x + 3 < 0$ and $f(x) = x^2 - 4x - 3$, then

(A) $0 < f(x) < 18$

(B) $f(x) \geq \dfrac{3}{2}$

(C) $f(x) \geq 12$

(D) $f(x) \geq 0$

(E) $2 < f(x) < 18$

$x^2 + 4x + 3 < 0 \rightarrow (x + 3)(x + 1) < 0$

$\therefore -3 < x < -1$, and $f(x) = (x - 2)^2 - 7$

Using graphing utility, we have min $y = 2$ and max $y = 18$ in the given domain, $-3 < x < -1$.

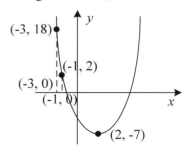

Ans. (E)

47. The length of the vector that could correctly be used to represent in the complex plane the number $z = 1 + 3i$ is

(A) 4
(B) $\sqrt{10}$
(C) 3
(D) $2\sqrt{5}$
(E) $\sqrt{11}$

The magnitude of the complex number $z = a + bi$ form is
$$|z| = \sqrt{a^2 + b^2}$$
Therefore, $|z| = |1 + 3i| = \sqrt{1^2 + 3^2} = \sqrt{10}$.

Ans. (B)

48. If A and B are different points in a space, the set of all points in this space that are closer to A than to B is

(A) the region of the space on one side of a plane
(B) the interior of a sphere
(C) a wedge-shaped region of the plane
(D) the region of the space bounded by a cube
(E) the interior of a circle

(Perpendicular Bisector Plane)
Referring to the fact that the P.B. line is the set of all points which is the equidistance from the point A and B, any point on the side of A from this P.B. line will be closer to B.

Ans. (A)

49. Which of the following lines are asymptotes of the graph of $y = \dfrac{x^2+x+1}{x}$?

I. $x = 0$
II. $y = 0$
III. $y = x + 1$

(A) I only
(B) II only
(C) I and II only
(D) I and III only
(E) I, II, and III

$$y = \frac{x^2+x+1}{x} = \frac{x^2}{x} + \frac{x}{x} + \frac{1}{x} = x + 1 + \frac{1}{x} = (x + 1) + \frac{1}{x}$$
∴ we have $(x + 1)$ as a slant asymptotes and $x = 0$ as a vertical asymptote.
∴ (I) and (III)

Ans. (D)

50. The radius of the base of a right circular cone is 5 and the radius of a parallel cross section is 3. If the distance between the base and the cross section is 6, what is the volume of the cone?

(A) 57
(B) 85
(C) 157
(D) 236
(E) 393

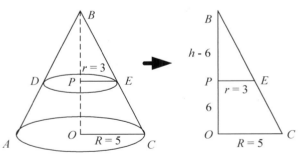

Now, let $\overline{OB} = h$, then $\overline{PB} = h - 6$.
∴ Using the property of similar triangles, we get $\dfrac{h-6}{3} = \dfrac{h}{5}$.
∴ $3h = 5h - 30$
∴ $2h = 30$
∴ $h = 15$
∴ Volume cone $= \dfrac{1}{3}\pi R^2 h = \dfrac{1}{3}\pi(5)^2(15) = 125\pi = 392.7$

Ans. (E)

Model Test No. 05

---

50 Questions / 60 Minutes

Directions: For each question, determine which of the answer choices is correct and fill in the oval on the answer sheet that corresponds to your choice.

Notes:

1. You will need to use a scientific or graphing calculator to answer some of the questions.
2. Be sure your calculator is in degree mode.
3. Each figure on this test is drawn as accurately as possible unless it is specifically indicated that the figure has not been drawn to scale.
4. The domain of any function f is the set of all real numbers x for which $f(x)$ is also a real number, unless the question indicates that the domain has been restricted in some way.
5. The box below contains five formulas that you may need to answer one or more of the questions.

REFERENCE INFORMATION

THE FOLLOWING INFORMATION IS FOR YOUR REFERENCE IN ANSWERING SOME OF THE QUESTIONS IN THIS TEST.

Volume of a right circular cone with radius r and height h: $V = \dfrac{1}{3}\pi r^2 h$

Lateral Area of a right circular cone with circumference of the base c and slant height l: $S = \dfrac{1}{2}cl$

Volume of a sphere with radius r: $V = \dfrac{4}{3}\pi r^3$

Surface Area of a sphere with radius r: $S = 4\pi r^2$

Volume of a pyramid with base area B and height h: $V = \dfrac{1}{3}Bh$

Answer Sheet
Model Test No. 05

1 Ⓐ Ⓑ Ⓒ Ⓓ Ⓔ	14 Ⓐ Ⓑ Ⓒ Ⓓ Ⓔ	27 Ⓐ Ⓑ Ⓒ Ⓓ Ⓔ	40 Ⓐ Ⓑ Ⓒ Ⓓ Ⓔ
2 Ⓐ Ⓑ Ⓒ Ⓓ Ⓔ	15 Ⓐ Ⓑ Ⓒ Ⓓ Ⓔ	28 Ⓐ Ⓑ Ⓒ Ⓓ Ⓔ	41 Ⓐ Ⓑ Ⓒ Ⓓ Ⓔ
3 Ⓐ Ⓑ Ⓒ Ⓓ Ⓔ	16 Ⓐ Ⓑ Ⓒ Ⓓ Ⓔ	29 Ⓐ Ⓑ Ⓒ Ⓓ Ⓔ	42 Ⓐ Ⓑ Ⓒ Ⓓ Ⓔ
4 Ⓐ Ⓑ Ⓒ Ⓓ Ⓔ	17 Ⓐ Ⓑ Ⓒ Ⓓ Ⓔ	30 Ⓐ Ⓑ Ⓒ Ⓓ Ⓔ	43 Ⓐ Ⓑ Ⓒ Ⓓ Ⓔ
5 Ⓐ Ⓑ Ⓒ Ⓓ Ⓔ	18 Ⓐ Ⓑ Ⓒ Ⓓ Ⓔ	31 Ⓐ Ⓑ Ⓒ Ⓓ Ⓔ	44 Ⓐ Ⓑ Ⓒ Ⓓ Ⓔ
6 Ⓐ Ⓑ Ⓒ Ⓓ Ⓔ	19 Ⓐ Ⓑ Ⓒ Ⓓ Ⓔ	32 Ⓐ Ⓑ Ⓒ Ⓓ Ⓔ	45 Ⓐ Ⓑ Ⓒ Ⓓ Ⓔ
7 Ⓐ Ⓑ Ⓒ Ⓓ Ⓔ	20 Ⓐ Ⓑ Ⓒ Ⓓ Ⓔ	33 Ⓐ Ⓑ Ⓒ Ⓓ Ⓔ	46 Ⓐ Ⓑ Ⓒ Ⓓ Ⓔ
8 Ⓐ Ⓑ Ⓒ Ⓓ Ⓔ	21 Ⓐ Ⓑ Ⓒ Ⓓ Ⓔ	34 Ⓐ Ⓑ Ⓒ Ⓓ Ⓔ	47 Ⓐ Ⓑ Ⓒ Ⓓ Ⓔ
9 Ⓐ Ⓑ Ⓒ Ⓓ Ⓔ	22 Ⓐ Ⓑ Ⓒ Ⓓ Ⓔ	35 Ⓐ Ⓑ Ⓒ Ⓓ Ⓔ	48 Ⓐ Ⓑ Ⓒ Ⓓ Ⓔ
10 Ⓐ Ⓑ Ⓒ Ⓓ Ⓔ	23 Ⓐ Ⓑ Ⓒ Ⓓ Ⓔ	36 Ⓐ Ⓑ Ⓒ Ⓓ Ⓔ	49 Ⓐ Ⓑ Ⓒ Ⓓ Ⓔ
11 Ⓐ Ⓑ Ⓒ Ⓓ Ⓔ	24 Ⓐ Ⓑ Ⓒ Ⓓ Ⓔ	37 Ⓐ Ⓑ Ⓒ Ⓓ Ⓔ	50 Ⓐ Ⓑ Ⓒ Ⓓ Ⓔ
12 Ⓐ Ⓑ Ⓒ Ⓓ Ⓔ	25 Ⓐ Ⓑ Ⓒ Ⓓ Ⓔ	38 Ⓐ Ⓑ Ⓒ Ⓓ Ⓔ	
13 Ⓐ Ⓑ Ⓒ Ⓓ Ⓔ	26 Ⓐ Ⓑ Ⓒ Ⓓ Ⓔ	39 Ⓐ Ⓑ Ⓒ Ⓓ Ⓔ	

USE THIS SPACE FOR SCRATCH WORK

1. What is the value of $\sqrt{x}$, if $x = 11 + 4\sqrt{6}$?

   (A) $3 + 2\sqrt{6}$
   (B) $3 - 2\sqrt{6}$
   (C) $\sqrt{3} - 2\sqrt{2}$
   (D) $2\sqrt{2} + \sqrt{3}$
   (E) $2\sqrt{2} - \sqrt{3}$

2. If 4 and -3 are both zeros of the polynomial $p(x)$, then a factor of $p(x)$ is

   (A) $x^2 - 6$
   (B) $x^2 - x - 12$
   (C) $x^2 + 6$
   (D) $x^2 + x - 12$
   (E) $x^2 + x + 12$

3. What is the volume of a sphere, with center at $(1, 2, 3)$, that passes through point $(1,1,1)$?

   (A) 33.2
   (B) 46.8
   (C) 30
   (D) 28.4
   (E) 53.8

4. Two cities are respectively 3,200 miles and 4,800 miles distant from San Francisco, and the distance between the two cities is 5,000 miles. What is the angle between the two cities formed from the city of San Francisco?

   (A) 65°
   (B) 38°
   (C) 74°
   (D) 89°
   (E) 122°

5. A boy walks diagonally across a square lot. What percent does he save by not walking along the edges (approximately)?

   (A) 22
   (B) 29
   (C) 33
   (D) 20
   (E) 24

6. Find the value of $\log_7 \sqrt[3]{49}$

   (A) $\dfrac{2}{3}$

   (B) $-\dfrac{2}{9}$

   (C) $-\dfrac{4}{3}$

   (D) $\dfrac{3}{4}$

   (E) $-\dfrac{3}{2}$

7. What is the remainder of
   $$P(x) = \frac{2}{3}x^5 + \frac{1}{6}x^4 + x^3 + \frac{5}{12}x^2 + 1,$$

   when $P(x)$ is divided by $(x - 1)$?

   (A) 3.12
   (B) 4.75
   (C) 3.58
   (D) 4.50
   (E) 3.25

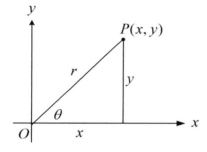

Figure 2

8. In Figure 2, $r \cos \theta + r \sin \theta =$

   (A) x
   (B) y
   (C) r
   (D) $x + y$
   (E) $r + y$

9. Solve the equation $x + 2\sqrt{x} - 3 = 0$.

   (A) 1 and $\dfrac{1}{4}$

   (B) $\dfrac{1}{4}$

   (C) 1
   (D) 0
   (E) 4

166

10. If $f(x) = 5^x$, then $f^{-1}(x) =$

(A) $\log_5 x$ for $x > 0$
(B) $\log_x 5$ for $x > 0$

(C) $\dfrac{x}{5}$ for $x > 0$

(D) $\dfrac{5}{x}$ for $x > 0$

(E) x^5 for $x > 0$

11. Find the radius of the circle whose equation is
$x^2 + y^2 + 24x - 10y = 0$

(A) 11
(B) 12
(C) 13
(D) 14
(E) 15

12. The converse of $\sim p \to q$ is equivalent to

(A) $p \to \sim q$
(B) $q \to p$
(C) $q \to \sim p$
(D) $p \to q$
(E) $\sim q \to p$

13. What is the sum of the infinite geometric series
$1 + \dfrac{1}{2} + \dfrac{1}{4} + \dfrac{1}{8} + \dfrac{1}{16} + \dfrac{1}{32} + \ldots$?

(A) $\dfrac{1}{2}$

(B) 1

(C) $\dfrac{3}{2}$

(D) 2

(E) $\dfrac{5}{2}$

14. A pyramid is cut by a plane parallel to its base at a distance from the base equal to one-thirds the length of the altitude. The area of the base is 18.
Find the area of the section determined by the pyramid and the cutting plane.

(A) 1
(B) 2
(C) 3
(D) 6
(E) 8

15. $\dfrac{1}{\sin\theta \cdot \cos\theta} - \tan\theta =$

(A) $\sin\theta$
(B) $\cos\theta$
(C) $\tan\theta$
(D) $\cot\theta$
(E) $\sec\theta$

16. If $f(x) = |2x - 3|$, then $f(3) =$

(A) $f\left(-\dfrac{3}{2}\right)$

(B) $f(-1)$

(C) $f(0)$

(D) $f\left(\dfrac{2}{3}\right)$

(E) $f\left(\dfrac{3}{2}\right)$

17. Given a point P, whose rectangular coordinates are (x, y), for which $x = \dfrac{a}{c}$ and $y = \dfrac{b}{c}$, what is the ratio of y to x ?

(A) $\dfrac{a}{c}$

(B) $\dfrac{a}{b}$

(C) $\dfrac{b}{a}$

(D) $\dfrac{c}{a}$

(E) 1

18. Suppose a car is parked at 80 meters north of you, and a patrol car is at 100 meter of northeastern corner from you, with an angle of 30° between them. What is the distance, in meters, between them?

(A) 133.2
(B) 115.1
(C) 50.4
(D) 66.3
(E) 52.4

19. $\dfrac{(n-1)!^4}{(n!(n-2)!)^2} =$

   (A) $\dfrac{1}{n}$

   (B) $\dfrac{1}{n^2}$

   (C) $\dfrac{n-1}{n}$

   (D) $\left(\dfrac{n-1}{n}\right)^2$

   (E) $(n-1)^2$

20. What is the *x*-intercept of the line containing the points (9, -5.5) and (-3, 4.5)?

   (A) 0.17
   (B) 0.83
   (C) 1.14
   (D) 2.40
   (E) 6

21. $\dfrac{2}{x} + y = 4$ and $x + \dfrac{2}{y} = 3$, then the ratio of *x* to *y* is

   (A) 1:2
   (B) 2:3
   (C) 3:1
   (D) 3:2
   (E) 3:4

22. A value that satisfies the equation
   $\sin^2 x + 4 \sin x = 0$ is (in degrees)

   (A) 0
   (B) 30
   (C) 60
   (D) 90
   (E) none of these

23. What is the largest rod that can just fit into a box
   8" × 12" × 9" (in inches)?

   (A) 14
   (B) 16
   (C) 17
   (D) 18
   (E) 19

24. What is the smallest positive angle that will make $3 + 2 \sin (x + \frac{\pi}{6})$ a minimum?

(A) 1.05
(B) 2.09
(C) 1.57
(D) 4.19
(E) 5.24

25. Which of the following is the solution set for $(-)(x + 1)(x - 2)(x - 3) < 0$?

(A) $x < -1$
(B) $-1 < x < 3$
(C) $-1 < x < 3$ or $x > 3$
(D) $x < -1$ or $2 < x < 3$
(E) $-1 < x < 2$ or $x > 3$

26. If $f(x) = 3x^2 - 5x - 3$, then $\frac{f(x+a)-f(x)}{a} =$

(A) $3a^2 - 5a - 3$
(B) $3x^2a^2 - 5xa - 3$
(C) $6x - 5 + 3a$
(D) $6x - 5$
(E) none of the above

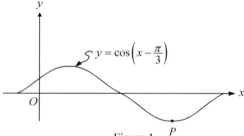

Figure 1

Note: Figure not drawn to scale

27. For $0 \leq x \leq 2\pi$, if the minimum value of the function $y = \cos (x - \frac{\pi}{3})$ occurs at point P, then what are the coordinates of P ?

(A) $(\frac{4\pi}{3}, -\pi)$

(B) $(\frac{4\pi}{3}, -1)$

(C) $(\frac{3\pi}{2}, -\pi)$

(D) $(\frac{3\pi}{2}, -1)$

(E) $(\frac{3\pi}{2}, 0)$

28. In how many ways can a committee of 1 woman and 3 men be selected from five men and four women so as to always include a particular man?

(A) 84
(B) 70
(C) 48
(D) 24
(E) 126

29. If $x_1 = \sqrt{3}$ and $x_{n+1} = \sqrt{3 + 2x_n}$, then $x_4 =$

(A) 1.732
(B) 2.542
(C) 2.843
(D) 2.947
(E) 2.982

30. Suppose the graph of $f(x) = x^2 - 4x + 7$ is translated 2 units left and 3 unit down. If the resulting graph represents $g(x)$, what is the value of $g(-0.5)$?

(A) 26.25
(B) 20.25
(C) 0.25
(D) -0.25
(E) -20.25

31. $2 \tan x - 2 \sin^2 x \cdot \tan x$ equals which one of the following?

(A) $\cos 2x$
(B) $2 \sin x$
(C) $2 \cos x$
(D) $\cos^2 x$
(E) $\sin 2x$

32. The Fibonacci sequence can be defined recursively as

$a_1 = 1$
$a_2 = 1$
$a_n = a_{n-1} + a_{n-2}$ for $n \geq 3$.

What is the 12th term of this sequence?

(A) 12
(B) 36
(C) 72
(D) 108
(E) 144

171

33. How many odd numbers greater than 30,000 may be formed using the digits 1, 2, 3, 4, and 5 if each digit must be used exactly once in each number?

(A) 42
(B) 48
(C) 64
(D) 96
(E) 112

34. If matrix A has dimensions $m \times n$ and matrix B has dimensions $n \times m$, where m and n are distinct positive integers, which of the following statements must be true?

I.   The product BA does not exist.
II. The product AB exists and has dimensions $m \times m$.
III. The product AB exists and has dimensions $n \times n$.

(A) I only
(B) II only
(C) III only
(D) I and II
(E) I and III

35. When drawn on the same set of axes, the graphs of $x^2 - y^2 = 4$ and $4(x - 1)^2 + y^2 = 25$ have in common exactly

(A) 0 points
(B) 1 point
(C) 2 points
(D) 3 points
(E) 4 points

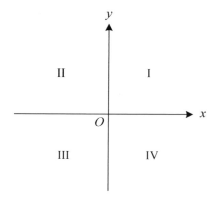

36. If $\sin \theta \cdot \cos \theta < 0$ and $\sin \theta \cdot \tan \theta > 0$, then θ must be in which quadrant in the figure above?

(A) I
(B) II
(C) III
(D) IV
(E) There is no quadrant in which both conditions are true.

USE THIS SPACE FOR SCRATCH WORK

37. If $f(-x) = -f(x)$ for all real numbers x and if (a, b) is a point on the graph of f, which of the following points must also be on the graph of f?

(A) $(-b, -a)$
(B) $(-a, -b)$
(C) $(-a, b)$
(D) $(a, -b)$
(E) (b, a)

38. Lines $\overline{AB}$ and $\overline{AC}$ are tangents to a circle at points B and C respectively. Minor arc BC is 2.5π inches, and the diameter of the circle is 12 inches. What is the number of degrees in angle BAC?

(A) 90°
(B) 95°
(C) 70°
(D) 100°
(E) 105°

39. The points in the rectangular coordinate plane are transformed in such a way that each point $P(x, y)$ is moved to the point $P'(3x, 3y)$. If the distance between a point P and the origin is d, then the distance between the points P and P' is

(A) $\frac{1}{d}$
(B) $\frac{d}{2}$
(C) d
(D) $2d$
(E) $3d$

40. The portion of the plane, whose equation is $12x + 15y + 10z = 120$, that lies in the first octant forms a pyramid with the coordinate planes. Find its volume.

(A) 90
(B) 120
(C) 140
(D) 150
(E) 160

41. Which of the following shifts of the graph of $y = 2x^2$ would result in the graph of $y = 2x^2 - 4kx + 2k^2 + k$, where k is a constant greater than 2?

(A) Left 2 units and up k units
(B) Left k unit and up $k + 1$ units
(C) Right k unit and down $k + 1$ units
(D) Left k unit and up k units
(E) Right k unit and up k units

42. For what positive value of a will the line $y = ax - 5$ be tangent to the circle $x^2 + y^2 = 4$?

(A) 1

(B) $\dfrac{\sqrt{21}}{2}$

(C) $\dfrac{\sqrt{21}}{3}$

(D) $\dfrac{\sqrt{31}}{4}$

(E) 2

43. $(x-1)^3 - 2(x-1)^2 - (x-1) + 2 =$

(A) $x(x-2)(x-3)$
(B) $(x-1)(x-2)(x-3)$
(C) $x(x-1)(x-2)$
(D) $(x+1)(x+2)(x+3)$
(E) $x(x-1)(x-3)$

44. When a certain radioactive element decays, the amount that exists at any time t can be calculated by the function $A(t) = A_0 e^{\frac{-t}{1000}}$, where A_0 is the initial amount and t is the elapsed time in years. How many years would it take for an initial amount of 800 milligrams of this element to decay to 200 milligrams?

(A) 0.5
(B) 500
(C) 1,386
(D) 1,443
(E) 5,704

45. If the values of the function $g(x)$ represent the slope of the line tangent to the graph of the function $f(x)$, shown below, at each point (x, y), which of the following could be the graph of $g(x)$?

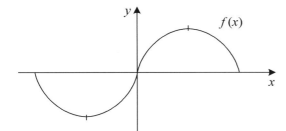

(A)
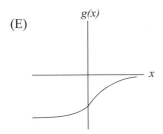

(B)

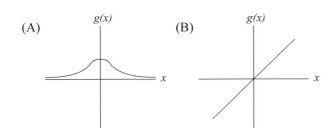

(C)

(D)

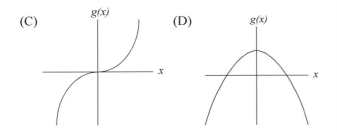

(E)

46. In a cube, what is the approximate angle θ formed by the longest diagonal to a diagonal of a base?

(A) 30°
(B) 250°
(C) 35°
(D) 45°
(E) 72°

175

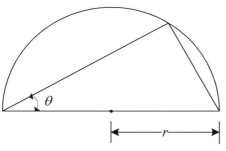

Figure 8

47. Figure 8 shows a triangle inscribed in a semicircle.
    If the area of the triangle is $8 \cdot \sin \theta \cdot \cos \theta$,
    what is the length of the radius of the semicircle?

    (A) 1
    (B) 2
    (C) 3
    (D) 4
    (E) 5

48. What is the set of all points in space that is equidistance
    from any two fixed points?

    (A) a triangle
    (B) a perpendicular bisector plane
    (C) a circle
    (D) a line
    (E) a sphere

49. The graph of $xy - x - y + 2 = 0$ can be
    expressed as a set of parametric equations.
    If $y = \dfrac{t}{t+1}$ and $x = f(t)$, then $f(t) =$

    (A) $t + 1$
    (B) $t - 1$
    (C) $t + 2$
    (D) $t - 2$
    (E) $\dfrac{t-3}{2}$

50. Given the following parametric equations:
    $x = 2\cos \theta + 1,\ y = 3\sin \theta$,
    which one of the following curve represents
    the above equations in rectangular form?

    (A) a circle
    (B) a parabola
    (C) a hyperbola
    (D) an ellipse
    (E) a line

176

Answer Key
Model Test No. 05

1	D	14	E	27	B	40	E
2	B	15	D	28	D	41	E
3	B	16	C	29	D	42	B
4	C	17	C	30	C	43	A
5	B	18	C	31	E	44	C
6	A	19	D	32	E	45	D
7	E	20	D	33	A	46	C
8	D	21	E	34	B	47	B
9	C	22	A	35	C	48	B
10	A	23	C	36	D	49	C
11	C	24	D	37	B	50	D
12	C	25	E	38	E		
13	D	26	C	39	D		

How to Score the SAT Subject Test in Mathematics Level 2

When you take an actual SAT Subject Test in Mathematics Level 2, your answer sheet will be "read" by a scanning machine that will record your responses to each question. Then a computer will compare your answers with the correct answers and produce your raw score. You get one point for each correct answer. For each wrong answer, you lose one-fourth of a point. Questions you omit (and any for which you mark more than one answer) are not counted. This raw score is converted to a scaled score that is reported to you and to the colleges you specify.

Finding Your Raw Test Score

STEP 1: Table A lists the correct answers for all the questions on the Subject Test in Mathematics Level 2 that is reproduced in this book. It also serves as a worksheet for you to calculate your raw score.

- Compare your answers with those given in the table.
- Put a check in the column marked "Right" if your answer is correct.
- Put a check in the column marked "Wrong" if your answer is incorrect.
- Leave both columns blank if you omitted the question.

STEP 2: Count the number of right answers.
Enter the total here: _____

STEP 3: Count the number of wrong answers.
Enter the total here: _____

STEP 4: Multiply the number of wrong answers by .250.
Enter the product here: _____

STEP 5: Subtract the result obtained in Step 4 from the total you obtained in Step 2.
Enter the result here: _____

STEP 6: Round the number obtained in Step 5 to the nearest whole number.
Enter the result here: _____

The number you obtained in Step 6 is your raw score.

Scaled Score Conversion Table
Subject Test in Mathematics Level 2

Raw Score	Scaled Score	Raw Score	Scaled Score	Raw Score	Scaled Score
50	800	28	630	6	470
49	800	27	630	5	460
48	800	26	620	4	450
47	800	25	610	3	440
46	800	24	600	2	430
45	800	23	600	1	420
44	800	22	590	0	410
43	790	21	580	-1	400
42	780	20	580	-2	390
41	770	19	570	-3	370
40	760	18	560	-4	360
39	750	17	560	-5	350
38	740	16	550	-6	340
37	730	15	540	-7	340
36	710	14	530	-8	330
35	700	13	530	-9	330
34	690	12	520	-10	320
33	680	11	510	-11	310
32	670	10	500	-12	300
31	660	9	490		
30	650	8	480		
29	640	7	480		

1. What is the value of $\sqrt{x}$, if $x = 11 + 4\sqrt{6}$?

    (A) $3 + 2\sqrt{6}$ (B) $3 - 2\sqrt{6}$
    (C) $\sqrt{3} - 2\sqrt{2}$ (D) $2\sqrt{2} + \sqrt{3}$
    (E) $2\sqrt{2} - \sqrt{3}$

$x = 11 + 4\sqrt{6}$

$\therefore \sqrt{x} = \sqrt{11 + 4\sqrt{6}} = \sqrt{(2\sqrt{2} + \sqrt{3})^2} = 2\sqrt{2} + \sqrt{3}$

But the best way is to use calculator!

Ans. (D)

2. If 4 and -3 are both zeros of the polynomial $p(x)$, then a factor of $p(x)$ is

    (A) $x^2 - 6$ (B) $x^2 - x - 12$
    (C) $x^2 + 6$ (D) $x^2 + x - 12$
    (E) $x^2 + x + 12$

4 and (-)3 are both zeros of the polynomial $p(x)$.
That means, by <u>Factor Theorem,</u>
we get $p(4) = 0$ and $p(-3) = 0$.
$\therefore p(x)$ must have $(x - 4)$ and $(x + 3)$ factors.
$\therefore p(x) = (x - 4)(x + 3) = x^2 - x - 12$

Ans. (B)

3. What is the volume of a sphere, with center at $(1, 2, 3)$, that passes through point $(1,1,1)$?

    (A) 33.2
    (B) 46.8
    (C) 30
    (D) 28.4
    (E) 53.8

Since the equation of the sphere with center at $(1, 2, 3)$ is, $(x - 1)^2 + (y - 2)^2 + (z - 3)^2 = r^2$, and this equation must satisfy $(1, 1, 1)$, we get $(1 - 1)^2 + (1 - 2)^2 + (1 - 3)^2 = r^2$
$\therefore 5 = r^2$
$\therefore r = \sqrt{5}$
But the volume of the sphere is, $V = \frac{4}{3}\pi r^3 = \frac{4}{3}\pi(\sqrt{5})^3 = 46.8$

Ans. (B)

4. Two cities are respectively 3,200 miles and 4,800 miles distant from San Francisco, and the distance between the two cities is 5,000 miles. What is the angle between the two cities formed from the city of San Francisco?

    (A) 65°
    (B) 38°
    (C) 74°
    (D) 89°
    (E) 122°

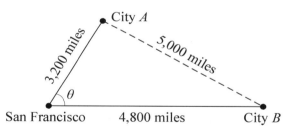

Refer to the figure, and also the Law of Cosine,
we get $5,000^2 = 3,200^2 + 4,800^2 - 2(3,200)(4,800) \cdot \cos\theta$
$\therefore \theta = \cos^{-1}(0.27) = 74°$

Ans. (C)

5. A boy walks diagonally across a square lot. What percent does he save by not walking along the edges (approximately)?

    (A) 22
    (B) 29
    (C) 33
    (D) 20
    (E) 24

The distance of walking edges will be $(1 + 1) = 2$, while the distance of walking diagonally will be $\sqrt{2}$.
$\therefore \frac{2-\sqrt{2}}{2} \times 100\% = 29.3\%$

Ans. (B)

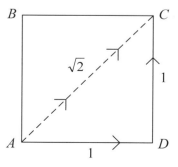

6. Find the value of $\log_7 \sqrt[3]{49}$

(A) $\dfrac{2}{3}$　(B) $-\dfrac{2}{9}$　(C) $-\dfrac{4}{3}$　(D) $\dfrac{3}{4}$　(E) $-\dfrac{3}{2}$

Refer to the properties of Log function.

We have, $\log x^m = m \cdot \log x$, and $\log_y x = \dfrac{\log x}{\log y} = \dfrac{\ln x}{\ln y}$.

$\therefore \log_7 \sqrt[3]{49} = \log_7 49^{\frac{1}{3}} = \log_7 (7^2)^{\frac{1}{3}}$

$= \log_7 7^{\frac{2}{3}} = \dfrac{2}{3} \log_7 7 = \dfrac{2}{3}(1) = \dfrac{2}{3}$.

Ans. (A)

7. What is the remainder of
$P(x) = \dfrac{2}{3}x^5 + \dfrac{1}{6}x^4 + x^3 + \dfrac{5}{12}x^2 + 1$,
when $P(x)$ is divided by $(x - 1)$?

(A) 3.12
(B) 4.75
(C) 3.58
(D) 4.50
(E) 3.25

Referring to the Remainder Theorem in our previous note, our remainder becomes $P(1)$.

$\therefore P(1) = \dfrac{2}{3} + \dfrac{1}{6} + 1 + \dfrac{5}{12} + 1 = \dfrac{39}{12} = \dfrac{13}{4}$

Ans. (E)

(Please note that $P(x) = (x - a)Q(x) + R(x)$.
$\therefore P(a) = (a - a)Q(a) + R(a) = 0 + R(a)$,
which is the remainder part.)

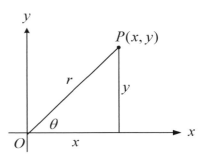

Figure 2

8. In Figure 2, $r \cos \theta + r \sin \theta =$

(A) x　(B) y　(C) r　(D) $x + y$　(E) $r + y$

In polar coordinates, we get $x = r \cos \theta$, $y = r \sin \theta$
where $r = \sqrt{x^2 + y^2}$, and $\theta = \tan^{-1} \dfrac{1}{x}$.
$\therefore r \cos \theta + r \sin \theta = x + y$

Ans. (D)

9. Solve the equation $x + 2\sqrt{x} - 3 = 0$.

(A) 1 and $\dfrac{1}{4}$　(B) $\dfrac{1}{4}$　(C) 1

(D) 0　(E) 4

$x + 2\sqrt{x} - 3 = 0$, now, let $\sqrt{x} = u$, $u > 0$
Then, $u^2 + 2u - 3 = 0$
$\therefore (u + 3)(u - 1) = 0$
$\therefore u = -3$ or $u = 1$, but $u > 0$
$\therefore u = 1$ only
$\therefore u = \sqrt{x} = 1$
$\therefore x = 1$

Ans. (C)

10. If $f(x) = 5^x$, then $f^{-1}(x) =$

(A) $\log_5 x$ for $x > 0$　(B) $\log_x 5$ for $x > 0$

(C) $\dfrac{x}{5}$ for $x > 0$　(D) $\dfrac{5}{x}$ for $x > 0$

(E) x^5 for $x > 0$

$y = f(x) = 5^x$, $y > 0$
Therefore, the inverse function of $y = 5^x$
becomes $x = 5^y$ by switching x and y.
Now, by applying $\log_5$ on both side,
we get $\log_5 x = \log_5 5^y = y \cdot \log_5 5 = y(1) = y$
$\therefore y = \log_5 x$, $x > 0$

Ans. (A)

11. Find the radius of the circle whose equation is
$x^2 + y^2 + 24x - 10y = 0$

(A) 11   (B) 12   (C) 13   (D) 14   (E) 15

$x^2 + y^2 + 24x - 10y = x^2 + 24x + y^2 - 10y = 0,$
or $(x + 12)^2 + (y - 5)^2 = 0 + 12^2 + 5^2 = 169$
∴ $(x + 12)^2 + (y - 5)^2 = 13^2$
∴ $r = 13$ with center $(-12, 5)$

Ans. (C)

12. The converse of $\sim p \rightarrow q$ is equivalent to

(A) $p \rightarrow \sim q$ (B) $q \rightarrow p$ (C) $q \rightarrow \sim p$
(D) $p \rightarrow q$ (E) $\sim q \rightarrow p$

In our previous note on <u>Logic</u>, the converse of $\sim p \rightarrow q$ is by switching the order of the statement as $q \rightarrow \sim p$.

Ans. (C)

13. What is the sum of the infinite geometric series
$1 + \dfrac{1}{2} + \dfrac{1}{4} + \dfrac{1}{8} + \dfrac{1}{16} + \dfrac{1}{32} + \ldots$?

(A) $\dfrac{1}{2}$ (B) 1   (C) $\dfrac{3}{2}$ (D) 2   (E) $\dfrac{5}{2}$

For a given Geometric Sequence; $a_1, a_2, a_3, \ldots, a_n, \ldots,$
we get $a_n = a_1 r^{n-1}$ and $s_n = a_1 \dfrac{1 - r^n}{1 - r}$.
Furthermore, for an infinite series,
we have the sum, $s = a_1 \dfrac{1}{1 - r}$, $|r| < 1$.
Here, since our $r = \dfrac{1}{2}$, and $|r| < 1$, we get $s = (1) \dfrac{1}{1 - \frac{1}{2}} = 2$

Ans. (D)

14. A pyramid is cut by a plane parallel to its base at a distance from the base equal to one-thirds the length of the altitude. The area of the base is 18.
Find the area of the section determined by the pyramid and the cutting plane.

(A) 1
(B) 2
(C) 3
(D) 6
(E) 8

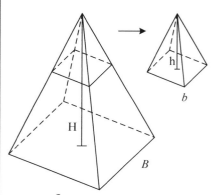

H:h = 1 : $\dfrac{2}{3}$

Since the ratio of its heights becomes $1 : \dfrac{2}{3}$, we get the ratio of

the base area B:b becomes $(1)^2 : \left(\dfrac{2}{3}\right)^2 = 1 : \dfrac{4}{9}$.

Here B = 18

∴ b = $18 \times \dfrac{4}{9} = 8$

Ans. (E)

15. $\dfrac{1}{\sin \theta \cdot \cos \theta} - \tan \theta =$

(A) $\sin \theta$ (B) $\cos \theta$ (C) $\tan \theta$
(D) $\cot \theta$ (E) $\sec \theta$

$\dfrac{1}{\sin \theta \cdot \cos \theta} - \tan \theta = \dfrac{1}{\sin \theta \cdot \cos \theta} - \dfrac{\sin \theta}{\cos \theta}$

$= \dfrac{1 - \sin^2 \theta}{\sin \theta \cdot \cos \theta} = \dfrac{\cos^2 \theta}{\sin \theta \cdot \cos \theta} = \dfrac{\cos \theta}{\sin \theta} = \cot \theta$

Ans. (D)

16. If $f(x) = |2x - 3|$, then $f(3) =$

   (A) $f\left(-\dfrac{3}{2}\right)$

   (B) $f(-1)$

   (C) $f(0)$

   (D) $f\left(\dfrac{2}{3}\right)$

   (E) $f\left(\dfrac{3}{2}\right)$

$f(3) = |2(3) - 3| = |3| = 3$
But $f(0) = |2(0) - 3| = |-3| = 3$

Ans. (C)

17. Given a point P, whose rectangular coordinates are (x, y), for which $x = \dfrac{a}{c}$ and $y = \dfrac{b}{c}$, what is the ratio of y to x ?

   (A) $\dfrac{a}{c}$ (B) $\dfrac{a}{b}$ (C) $\dfrac{b}{a}$ (D) $\dfrac{c}{a}$ (E) 1

Given $x = \dfrac{a}{c}$, $y = \dfrac{b}{c}$, we get $c = \dfrac{a}{x}$ and $c = \dfrac{b}{y}$.

$\therefore c = \dfrac{a}{x} = \dfrac{b}{y}$, or $\dfrac{y}{x} = \dfrac{b}{a}$

Ans. (C)

18. Suppose a car is parked at 80 meters north of you, and a patrol car is at 100 meter of northeastern corner from you, with an angle of 30° between them. What is the distance, in meters, between them?

   (A) 133.2
   (B) 115.1
   (C) 50.4
   (D) 66.3
   (E) 52.4

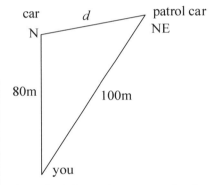

Given the information, we get the above figure. Then, the distance $d = \sqrt{100^2 + 80^2 - 2 \cdot 100 \cdot 80 \cdot \cos 30°} = 50.4$m

Ans. (C)

19. $\dfrac{(n-1)!^4}{(n!(n-2)!)^2} =$

   (A) $\dfrac{1}{n}$

   (B) $\dfrac{1}{n^2}$

   (C) $\dfrac{n-1}{n}$

   (D) $\left(\dfrac{n-1}{n}\right)^2$

   (E) $(n-1)^2$

$\dfrac{((n-1)!)^4}{(n!(n-2)!)^2} = \dfrac{((n-1)!(n-1)!)^2}{(n!(n-2)!)^2} = \left(\dfrac{(n-1)!}{n!}\right)^2 \left(\dfrac{(n-1)!}{(n-2)!}\right)^2$

$= \left(\dfrac{1}{n}\right)^2 \left(\dfrac{n-1}{1}\right)^2 = \left(\dfrac{n-1}{n}\right)^2$

Ans. (D)

(Note: the best way to solve is, just plug in $n = 3$)

183

20. What is the *x*-intercept of the line containing the points (9, -5.5) and (-3, 4.5)?

    (A) 0.17
    (B) 0.83
    (C) 1.14
    (D) 2.40
    (E) 6

For this problem, we need to review a linear function:
That is, for a given linear function, $y - y_1 = m(x - x_1)$,
we have slope $m = \dfrac{y_2 - y_1}{x_2 - x_1}$.

$\therefore$ With the two points (9, -5.5) and (-3, 4.5),
we get $m = \dfrac{4.5 - (-5.5)}{-3 - 9} = \dfrac{10}{-12} = (-)\dfrac{5}{6}$.

$\therefore y - (-5.5) = -\dfrac{5}{6}(x - 9)$

Since we need to get *x*-intercept, where $y = 0$,
we get $0 + 5.5 = (-)\dfrac{5}{6}(x - 9)$.

$\therefore 5.5 = -\dfrac{5}{6}x + \dfrac{15}{2}$ $\therefore \dfrac{5}{6}x = 2$ $\therefore x = \dfrac{12}{5} = 2.4$

Ans. (D)

---

21. $\dfrac{2}{x} + y = 4$ and $x + \dfrac{2}{y} = 3$, then the ratio of *x* to *y* is

    (A) 1:2
    (B) 2:3
    (C) 3:1
    (D) 3:2
    (E) 3:4

Let $\dfrac{2}{x} + y = 4$ be eq(1), and $x + \dfrac{2}{y} = 3$ be eq(2).
Then, multiplying *x* on both sides of eq(1),
we get $2 + xy = 4x$, eq(3).
And also multiplying *y* on both sides of eq(2),
we get $xy + 2 = 3y$, eq(4).
Now, eq(3) - eq(4) gives $0 = 4x - 3y$
$\therefore 4x = 3y$
$\therefore \dfrac{x}{y} = \dfrac{3}{4}$

Ans. (E)

---

22. A value that satisfies the equation $\sin^2 x + 4\sin x = 0$ is (in degrees)

    (A) 0   (B) 30   (C) 60
    (D) 90   (E) none of these

Using a graphing utility, let $y_1 = (\sin x)^2 + 4\sin x$,
we get $x = 0$, 180° … or, mathematically, $\sin x(\sin x + 4) = 0$
$\therefore \sin x = 0$, $\sin x \neq -4$
$\therefore x = 0°$, 180° …

Ans. (A)

---

23. What is the largest rod that can just fit into a box 8" × 12" × 9" (in inches)?

    (A) 14   (B) 16   (C) 17   (D) 18   (E) 19

The longest diagonal, $D = \sqrt{8^2 + 12^2 + 9^2} = 17$

Ans. (C)

---

24. What is the smallest positive angle that will make $3 + 2\sin\left(x + \dfrac{\pi}{6}\right)$ a minimum?

    (A) 1.05
    (B) 2.09
    (C) 1.57
    (D) 4.19
    (E) 5.24

Let $y_1 = 3 + 2\sin\left(x + \dfrac{\pi}{6}\right)$ in graphing mode,
we see that min occur at $x = 4.19$ in radian mode.
Or, mathematically, to get the minimum $2\sin\left(x + \dfrac{\pi}{6}\right) = -2$,
because $-1 \leq \sin\left(x + \dfrac{\pi}{6}\right) \leq 1$.
$\therefore \sin\left(x + \dfrac{\pi}{6}\right) = -1$, or $x + \dfrac{\pi}{6} = \dfrac{3\pi}{2}$
$\therefore x = \dfrac{3\pi}{2} - \dfrac{\pi}{6} = \dfrac{9\pi}{6} - \dfrac{\pi}{6} = \dfrac{4\pi}{3} = 4.19$

Ans. (D)

25. Which of the following is the solution set for $(-)(x + 1)(x - 2)(x - 3) < 0$?

(A) $x < -1$
(B) $-1 < x < 3$
(C) $-1 < x < 3$ or $x > 3$
(D) $x < -1$ or $2 < x < 3$
(E) $-1 < x < 2$ or $x > 3$

Using a graph of polynomial function, we get $-1 < x < 2, x > 3$

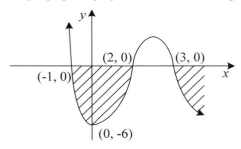

Ans. (E)

26. If $f(x) = 3x^2 - 5x - 3$, then $\dfrac{f(x+a)-f(x)}{a} =$

(A) $3a^2 - 5a - 3$
(B) $3x^2a^2 - 5xa - 3$
(C) $6x - 5 + 3a$
(D) $6x - 5$
(E) none of the above

Given $f(x)$ function,
we get $f(x + a) = 3(x + a)^2 - 5(x + a) - 3 =$
$3x^2 + 6ax + 3a^2 - 5x - 5a - 3$, eq(1).
$f(x) = 3x^2 - 5x - 3$, eq(2).
$\therefore \dfrac{\text{eq(1)}-\text{eq(2)}}{a} = \dfrac{6ax+3a^2-5a}{a} = 6x + 3a - 5$

Ans. (C)

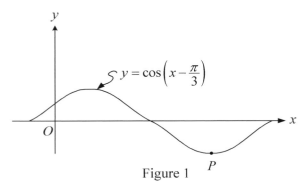

Figure 1

Note: Figure not drawn to scale

27. For $0 \le x \le 2\pi$, if the minimum value of the function $y = \cos (x - \dfrac{\pi}{3})$ occurs at point P,

then what are the coordinates of P ?

(A) $(\dfrac{4\pi}{3}, -\pi)$ (B) $(\dfrac{4\pi}{3}, -1)$ (C) $(\dfrac{3\pi}{2}, -\pi)$

(D) $(\dfrac{3\pi}{2}, -1)$ (E) $(\dfrac{3\pi}{2}, 0)$

As seen in the figure,
the minimum occurs when $\cos (x - \dfrac{\pi}{3}) = -1$,
but cosine function becomes $(-)1$ at $\theta = \pi$.
$\therefore x - \dfrac{\pi}{3} = \pi$

$\therefore x = \dfrac{4\pi}{3}$

Ans. (B)

28. In how many ways can a committee of 1 woman and 3 men be selected from five men and four women so as to always include a particular man?

(A) 84 (B) 70 (C) 48 (D) 24 (E) 126

Out of 5 men, a particular man is already fixed!
$\therefore$ We only have 4 men and 4 women.
Now, selecting 1 woman from 4 women becomes $_4C_1$,
and also 3 men from 4 men becomes $_4C_3$.
$\therefore {}_4C_2 \times {}_4C_1 = 24$

Ans. (D)

29. If $x_1 = \sqrt{3}$ and $x_{n+1} = \sqrt{3 + 2x_n}$, then $x_4 =$

(A) 1.732   (B) 2.542   (C) 2.843
(D) 2.947   (E) 2.982

$x_2 = \sqrt{3 + 2x_1} = \sqrt{3 + 2\sqrt{3}} = 2.54$
$x_3 = \sqrt{3 + 2x_2} = \sqrt{3 + 2(2.54)} = 2.84$
$x_4 = \sqrt{3 + 2x_3} = \sqrt{3 + 2(2.84)} = 2.947$

Ans. (D)

30. Suppose the graph of $f(x) = x^2 - 4x + 7$ is translated 2 units left and 3 unit down. If the resulting graph represents $g(x)$, what is the value of $g(-0.5)$?

(A) 26.25   (B) 20.25   (C) 0.25
(D) -0.25   (E) -20.25

Since x is translated 2 units left, we get, $x \to x - (-2) = x + 2$.
Also, y is 3 down, $y \to y - (-3) = y + 3$
$\therefore g(x) + 3 = f(x + 2)$, or $g(x) = f(x + 2) - 3$
$= (x + 2)^2 - 4(x + 2) + 7 - 3$
Now, $g\left(-\frac{1}{2}\right) = \left(\frac{3}{2}\right)^2 - 4\frac{3}{2} + 4 = \frac{9}{4} - 6 + 4 = \frac{1}{4}$

Ans. (C)

31. $2 \tan x - 2 \sin^2 x \cdot \tan x$ equals which one of the following?

(A) $\cos 2x$
(B) $2 \sin x$
(C) $2 \cos x$
(D) $\cos^2 x$
(E) $\sin 2x$

The best way to solve quickly is to use a graphing utility by letting, $y_1 = 2 \tan x - (2 \sin^2 x)(\tan x)$.
And then, try answer choice (A) $= y_2$, (B) $= y_3$, ... so on.
We will see answer choice (E) becomes the overlapping one.
Now, if we solve mathematically, $2 \tan x - 2 \sin^2 x \cdot \tan x$
$= 2\frac{\sin x}{\cos x} - 2(\sin^2 x)\frac{\sin x}{\cos x} = 2\frac{(\sin - \sin^3 x)}{\cos x}$
$= 2\frac{\sin x(1 - \sin^2 x)}{\cos x} = 2\frac{\sin x \cdot \cos^2 x}{\cos x} = 2 \sin x \cdot \cos x = \sin 2x$

Ans. (E)

32. The Fibonacci sequence can be defined recursively as

$a_1 = 1$
$a_2 = 1$
$a_n = a_{n-1} + a_{n-2}$ for $n \geq 3$.

What is the 12th term of this sequence?

(A) 12   (B) 36   (C) 72
(D) 108   (E) 144

$a_3 = a_2 + a_1 = 1 + 1 = 2$
$a_4 = a_3 + a_2 = 1 + 2 = 3$
$a_5 = a_4 + a_3 = 3 + 2 = 5$
$a_6 = a_5 + a_4 = 5 + 3 = 8$
...
But from this sequential pattern, we may extend the sequence as, 1, 1, 2, 3, 5, 8, 13, 21, 34, 55, 89, 144.
$\therefore$ The 12th term is 144

Ans. (E)

33. How many odd numbers greater than 30,000 may be formed using the digits 1, 2, 3, 4, and 5 if each digit must be used exactly once in each number?

(A) 42
(B) 48
(C) 64
(D) 96
(E) 112

1st, 2nd, 3rd, 4th, 5th
Now, to be greater than 30,000,
we could only have either 3, 4 or 5 on the 1st spot.
That is, for example, $\underline{3}$, _, _, _, _.
But we also need ODD number,
thus the last spot should be 1 or 5.
That means, $\underline{3}$, _, _, _, $\underline{1}$ or $\underline{3}$, _, _, _, $\underline{5}$ with
the possible choices of $3 \times 2 \times 1 = 6$ combination each!!
$\therefore 6 + 6 = 12$ combination for this case!!
We also have two other cases such as: $\underline{4}$, _, _, _, $\underline{1, 3}$ or $\underline{5}$,
or $3 \times 2 \times 1 \times (3) = 18$ combinations!!
Or, $\underline{5}$, _, _, _, $\underline{1}$ or $\underline{3}$ with $3 \times 2 \times 1 \times (2) = 12$ combinations.
Therefore, the total $= 12 + 18 + 12 = 42$.

Ans. (A)

34. If matrix A has dimensions $m \times n$ and matrix B has dimensions $n \times m$, where m and n are distinct positive integers, which of the following statements must be true?

I.   The product BA does not exist.
II. The product AB exists and has dimensions $m \times m$.
III. The product AB exists and has dimensions $n \times n$.

(A) I only
(B) II only
(C) III only
(D) I and II
(E) I and III

The product of the matrix must match the dimension of columns of the first matrix equal to the rows of the second matrix, such as $[p \times q][q \times r] = [p \times r]$.
Here, (I) $BA = [n \times m][m \times n] = [n \times n]$
$\therefore BA$ exist
$\therefore$ (I) is not true!
(II) $AB = [m \times n][n \times m] = [m \times m]$
$\therefore AB$ exist with $[m \times m]$
(III) Not true!

Ans. (B)

35. When drawn on the same set of axes, the graphs of $x^2 - y^2 = 4$ and $4(x-1)^2 + y^2 = 25$ have in common exactly

(A) 0 points
(B) 1 point
(C) 2 points
(D) 3 points
(E) 4 points

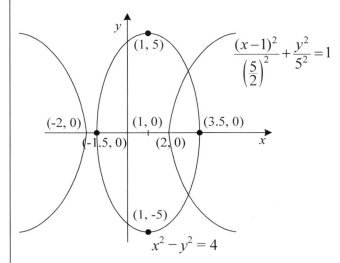

Ans. (C)

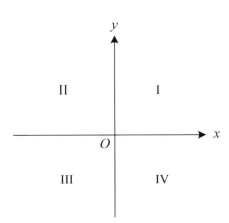

36. If $\sin \theta \cdot \cos \theta < 0$ and $\sin \theta \cdot \tan \theta > 0$, then θ must be in which quadrant in the figure above?

(A) I
(B) II
(C) III
(D) IV
(E) There is no quadrant in which both conditions are true.

$\sin \theta \cdot \cos \theta < 0$
$\therefore \sin \theta$ and $\cos \theta$ must have opposite signs!
$\therefore$ II or IV
$\sin \theta \cdot \tan \theta > 0$
$\therefore \sin \theta$ and $\tan \theta$ must have the same signs.
$\therefore$ I or IV
Now, IV is the only quadrant that satisfy both!!

Ans. (D)

37. If $f(-x) = -f(x)$ for all real numbers x and if (a, b) is a point on the graph of f, which of the following points must also be on the graph of f?

    (A) $(-b, -a)$ (B) $(-a, -b)$ (C) $(-a, b)$
    (D) $(a, -b)$ (E) (b, a)

The property $f(-x) = -f(x)$ is for ODD function, which is symmetric to origin $(0, 0)$, we must have (a, b) in quadrant I and also $(-a, -b)$ in quadrant III on the curve of $f(x)$.

Ans. (B)

38. Lines $\overline{AB}$ and $\overline{AC}$ are tangents to a circle at points B and C respectively. Minor arc BC is 2.5π inches, and the diameter of the circle is 12 inches. What is the number of degrees in angle BAC?

    (A) $90°$
    (B) $95°$
    (C) $70°$
    (D) $100°$
    (E) $105°$

$\overset{\frown}{BC} = 2.5\pi = r \cdot \theta$,
where $r = 6$
$\therefore \overset{\frown}{BC} = 2.5\pi = 6 \cdot \theta$

$\therefore \theta = \dfrac{2.5\pi}{6}$

$= \dfrac{2.5(180°)}{6} = 75°$

$\therefore \angle BAC$

$= 180° - 75° = 105°$

Ans. (E)

39. The points in the rectangular coordinate plane are transformed in such a way that each point $P(x, y)$ is moved to the point $P'(3x, 3y)$. If the distance between a point P and the origin is d, then the distance between the points P and P' is

    (A) $\dfrac{1}{d}$ (B) $\dfrac{d}{2}$ (C) d (D) $2d$ (E) $3d$

$\overline{PP'} = \sqrt{(3x - x)^2 + (3y - y)^2} = \sqrt{4x^2 + 4y^2}$
$= 2\sqrt{x^2 + y^2} = 2d$

Ans. (D)

40. The portion of the plane, whose equation is $12x + 15y + 10z = 120$, that lies in the first octant forms a pyramid with the coordinate planes. Find its volume.

    (A) 90
    (B) 120
    (C) 140
    (D) 150
    (E) 160

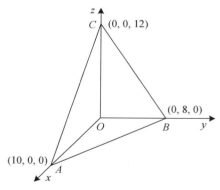

For the x-intercept, let $y = 0$, $z = 0$, then $x = \dfrac{120}{12} = 10$.
Similarly, we put $x = 0$ and $z = 0$, to get $15y = 120$
$\therefore y = \dfrac{120}{15} = 8$.
Also, $x = 0$ and $y = 0$ gives $10z = 120$
$\therefore z = 12$

The volume of pyramid becomes, $V = \dfrac{1}{3}$ (Base area) × height

$= \dfrac{1}{3} \times (\dfrac{1}{2} \times 10 \times 8) \times 12 = 160$

Ans. (E)

41. Which of the following shifts of the graph of $y = 2x^2$ would result in the graph of $y = 2x^2 - 4kx + 2k^2 + k$, where k is a constant greater than 2?

(A) Left 2 units and up k units
(B) Left k unit and up $k + 1$ units
(C) Right k unit and down $k + 1$ units
(D) Left k unit and up k units
(E) Right k unit and up k units

$y = 2x^2 - 4kx + 2k^2 + k = 2(x^2 - 2kx + k^2) + k$, or $y = 2(x - k)^2 + k \rightarrow$ this shows that $y = 2x^2$ has moved to right k units and up k units.

Ans. (E)

42. For what positive value of a will the line $y = ax - 5$ be tangent to the circle $x^2 + y^2 = 4$?

(A) 1

(B) $\dfrac{\sqrt{21}}{2}$

(C) $\dfrac{\sqrt{21}}{3}$

(D) $\dfrac{\sqrt{31}}{4}$

(E) 2

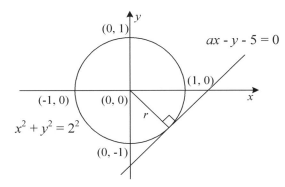

Referring to the distance formula from a point (x_1, y_1) to a line $Ax + By + C = 0$ is $d = \dfrac{|Ax_1 + By_1 + C|}{\sqrt{A^2 + B^2}}$, we have $(x_1, y_1) = (0, 0)$ and $d = r = 2$, and $A = a$, $B = -1$, $C = -5$

$\therefore d = r = 2 = \dfrac{|a(0) - (1)(0) + (-5)|}{\sqrt{a^2 + 1}}$

$\therefore 2 = \dfrac{5}{\sqrt{a^2 + 1}}$

$\therefore \sqrt{a^2 + 1} = \dfrac{5}{2}$, or $a^2 = \dfrac{21}{4}$

$\therefore a = \dfrac{\sqrt{21}}{2}$

Ans. (B)

43. $(x - 1)^3 - 2(x - 1)^2 - (x - 1) + 2 =$

(A) $x(x - 2)(x - 3)$ (B) $(x - 1)(x - 2)(x - 3)$
(C) $x(x - 1)(x - 2)$ (D) $(x + 1)(x + 2)(x + 3)$
(E) $x(x - 1)(x - 3)$

Let $A = x - 1$, then $A^3 - 2A^2 - A + 2 = A^2(A - 2) - (A - 2)$
$= (A - 2)(A^2 - 1) = (A - 2)(A + 1)(A - 1)$
Now, replacing $A = x - 1$, we get $(x - 3)(x)(x - 2)$.

Ans. (A)

44. When a certain radioactive element decays, the amount that exists at any time t can be calculated by the function $A(t) = A_0 e^{\frac{-t}{1000}}$, where A_0 is the initial amount and t is the elapsed time in years. How many years would it take for an initial amount of 800 milligrams of this element to decay to 200 milligrams?

(A) 0.5   (B) 500   (C) 1,386
(D) 1,443   (E) 5,704

$200 = 800 \cdot e^{\frac{-t}{1000}}$

$\therefore \dfrac{200}{800} = \dfrac{1}{4} = e^{\frac{-t}{1000}}$

Now, applying "ln" on both sides,
we get $\ln \dfrac{1}{4} = \ln e^{\frac{-t}{1000}} = \dfrac{-t}{1000} \ln e = \dfrac{-t}{1000}$ (1).

$\therefore t = -1000 \cdot \ln \dfrac{1}{4} = 1,386$

Ans. (C)

45. If the values of the function $g(x)$ represent the slope of the line tangent to the graph of the function $f(x)$, shown below, at each point (x, y), which of the following could be the graph of $g(x)$?

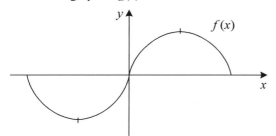

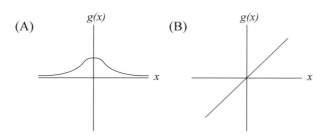

(A)   (B)

(C)   (D)

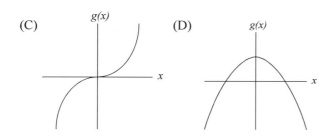

(E)

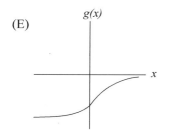

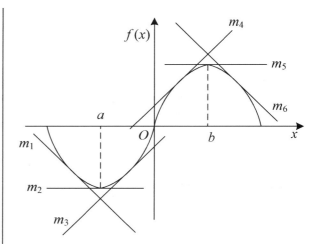

By examining the movement of the slopes:
$m_1 \rightarrow m_2 \rightarrow m_3 \rightarrow m_4 \rightarrow m_5 \rightarrow m_6$,
we get the graph of $g(x)$ is approximately similar to answer choice (D).

Ans. (D)

46. In a cube, what is the approximate angle θ formed by the longest diagonal to a diagonal of a base?

(A) 30°
(B) 250°
(C) 35°
(D) 45°
(E) 72°

Suppose the dimension of the cube is 1 x 1 x 1.

Then, the longest diagonal becomes $D = \sqrt{1^2 + 1^2 + 1^2} = \sqrt{3}$ and the diagonal of the base is $d = \sqrt{1^2 + 1^2} = \sqrt{2}$.

Therefore, the angle is, $\cos \theta = \dfrac{\sqrt{3}}{\sqrt{2}}$, and $\theta = \cos^{-1}\left(\dfrac{\sqrt{3}}{\sqrt{2}}\right) =$

35.3°

Ans. (C)

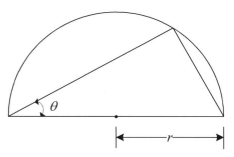

Figure 8

47. Figure 8 shows a triangle inscribed in a semicircle. If the area of the triangle is $8 \cdot \sin \theta \cdot \cos \theta$, what is the length of the radius of the semicircle?

(A) 1
(B) 2
(C) 3
(D) 4
(E) 5

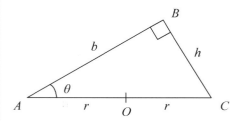

Since $\triangle ABC$ is inscribed in a semicircle, we have $\angle B = 90°$.

$\therefore \sin \theta = \dfrac{h}{2r}$, or $h = (2r) \sin \theta$.

Also, $\cos \theta = \dfrac{b}{2r}$, or $b = (2r) \cos \theta$.

$\therefore$ Area of $\triangle ABC = \dfrac{1}{2} bh = \dfrac{1}{2} (2r \cos \theta)(2r \sin \theta)$

$= 2r^2 \sin \theta \cdot \cos \theta = 8 \sin \theta \cdot \cos \theta$

$\therefore 2r^2 = 8$ or $r^2 = 4$

$\therefore r = 2$

Ans. (B)

48. What is the set of all points in space that is equidistance from any two fixed points?

(A) a triangle
(B) a perpendicular bisector plane
(C) a circle   (D) a line   (E) a sphere

The definition of a perpendicular bisector line in plane is the set of all points that is equidistance from any two fixed points. Therefore, in space, we have P.B. plane.

Ans. (B)

49. The graph of $xy - x - y + 2 = 0$ can be expressed as a set of parametric equations. If $y = \dfrac{t}{t+1}$ and $x = f(t)$, then $f(t) =$

(A) $t + 1$
(B) $t - 1$
(C) $t + 2$
(D) $t - 2$
(E) $\dfrac{t-3}{2}$

Replacing $x = f(t)$ and $y = \dfrac{t}{t+1}$,

$f(t) \cdot \left(\dfrac{t}{t+1}\right) - f(t) - \left(\dfrac{t}{t+1}\right) + 2 = 0$

$\therefore f(t) \cdot [\left(\dfrac{t}{t+1}\right) - 1] = \left(\dfrac{t}{t+1}\right) - 2$ or $f(t) \cdot \left(\dfrac{-1}{t+1}\right)$

$= \dfrac{t - 2(t+1)}{t+1} = \dfrac{t - 2t - 2}{t+1} = \dfrac{-t - 2}{t+1}$

$\therefore f(t) = \dfrac{-(t+2)}{t+1} \cdot \dfrac{t+1}{-1} = t + 2$

Ans. (C)

50. Given the following parametric equations: $x = 2\cos \theta + 1$, $y = 3\sin \theta$, which one of the following curve represents the above equations in rectangular form?

(A) a circle
(B) a parabola
(C) a hyperbola
(D) an ellipse
(E) a line

Since $\dfrac{x-1}{2} = \cos \theta$ and $\dfrac{y}{3} = \sin \theta$,

squaring both side of the above equations,

we get $\dfrac{(x-1)^2}{2^2} = \cos^2\theta$ and $\dfrac{y^2}{3^2} = \sin^2\theta$

$\therefore \dfrac{(x-1)^2}{2^2} + \dfrac{y^2}{3^2} = \cos^2\theta + \sin^2\theta = 1$, which is an ellipse with center $(1, 0)$ and major axis on y-axis with its length of 6.

Ans. (D)

Model Test No. 06

---

50 Questions / 60 Minutes

Directions: For each question, determine which of the answer choices is correct and fill in the oval on the answer sheet that corresponds to your choice.

Notes:

1. You will need to use a scientific or graphing calculator to answer some of the questions.
2. Be sure your calculator is in degree mode.
3. Each figure on this test is drawn as accurately as possible unless it is specifically indicated that the figure has not been drawn to scale.
4. The domain of any function f is the set of all real numbers x for which $f(x)$ is also a real number, unless the question indicates that the domain has been restricted in some way.
5. The box below contains five formulas that you may need to answer one or more of the questions.

REFERENCE INFORMATION

THE FOLLOWING INFORMATION IS FOR YOUR REFERENCE IN ANSWERING SOME OF THE QUESTIONS IN THIS TEST.

Volume of a right circular cone with radius r and height h: $V = \frac{1}{3}\pi r^2 h$

Lateral Area of a right circular cone with circumference of the base c and slant height l: $S = \frac{1}{2}cl$

Volume of a sphere with radius r: $V = \frac{4}{3}\pi r^3$

Surface Area of a sphere with radius r: $S = 4\pi r^2$

Volume of a pyramid with base area B and height h: $V = \frac{1}{3}Bh$

Answer Sheet
Model Test No. 06

1	ⒶⒷⒸⒹⒺ	14	ⒶⒷⒸⒹⒺ	27	ⒶⒷⒸⒹⒺ	40	ⒶⒷⒸⒹⒺ
2	ⒶⒷⒸⒹⒺ	15	ⒶⒷⒸⒹⒺ	28	ⒶⒷⒸⒹⒺ	41	ⒶⒷⒸⒹⒺ
3	ⒶⒷⒸⒹⒺ	16	ⒶⒷⒸⒹⒺ	29	ⒶⒷⒸⒹⒺ	42	ⒶⒷⒸⒹⒺ
4	ⒶⒷⒸⒹⒺ	17	ⒶⒷⒸⒹⒺ	30	ⒶⒷⒸⒹⒺ	43	ⒶⒷⒸⒹⒺ
5	ⒶⒷⒸⒹⒺ	18	ⒶⒷⒸⒹⒺ	31	ⒶⒷⒸⒹⒺ	44	ⒶⒷⒸⒹⒺ
6	ⒶⒷⒸⒹⒺ	19	ⒶⒷⒸⒹⒺ	32	ⒶⒷⒸⒹⒺ	45	ⒶⒷⒸⒹⒺ
7	ⒶⒷⒸⒹⒺ	20	ⒶⒷⒸⒹⒺ	33	ⒶⒷⒸⒹⒺ	46	ⒶⒷⒸⒹⒺ
8	ⒶⒷⒸⒹⒺ	21	ⒶⒷⒸⒹⒺ	34	ⒶⒷⒸⒹⒺ	47	ⒶⒷⒸⒹⒺ
9	ⒶⒷⒸⒹⒺ	22	ⒶⒷⒸⒹⒺ	35	ⒶⒷⒸⒹⒺ	48	ⒶⒷⒸⒹⒺ
10	ⒶⒷⒸⒹⒺ	23	ⒶⒷⒸⒹⒺ	36	ⒶⒷⒸⒹⒺ	49	ⒶⒷⒸⒹⒺ
11	ⒶⒷⒸⒹⒺ	24	ⒶⒷⒸⒹⒺ	37	ⒶⒷⒸⒹⒺ	50	ⒶⒷⒸⒹⒺ
12	ⒶⒷⒸⒹⒺ	25	ⒶⒷⒸⒹⒺ	38	ⒶⒷⒸⒹⒺ		
13	ⒶⒷⒸⒹⒺ	26	ⒶⒷⒸⒹⒺ	39	ⒶⒷⒸⒹⒺ		

1. If $8x + 12 = \dfrac{k}{3}(2x + 3)$ for all x, then $k =$

   (A) $\dfrac{1}{4}$

   (B) 3
   (C) 4
   (D) 12
   (E) 24

2. Which of the following is an equation of a line perpendicular to $y = 3x - 2$?

   (A) $y = 2x + 3$
   (B) $y = -3x + 2$

   (C) $y = -\dfrac{1}{3}x + 5$

   (D) $y = \dfrac{1}{3}x - 2$

   (E) $y = \dfrac{1}{-3x+2}$

3. Find the solution set of the inequality $x^2 - 3x - 4 < 0$.

   (A) $x > -1$
   (B) $-1 < x < 4$
   (C) $x > 4$ and $x < -1$
   (D) $x > 4$ or $x < -1$
   (E) $x < 4$

4. If $f(x, y) = (\ln \sqrt{2}x^3)e^{2\sqrt{y}}$,
   what is the approximate value of $f(\sqrt{2}, 2)$?

   (A) 23.45
   (B) 24.35
   (C) 25.34
   (D) 25.43
   (E) 27.25

5. $\dfrac{\sin\frac{2\pi}{3}\cos\frac{\pi}{3}}{\tan 45°} =$

(A) $\dfrac{\sqrt{3}}{2}$

(B) $-\dfrac{\sqrt{3}}{4}$

(C) $\dfrac{\sqrt{6}}{4}$

(D) $-\dfrac{\sqrt{3}}{2}$

(E) $\dfrac{\sqrt{3}}{4}$

6. The graph of the rational function f, where $f(x) = \dfrac{x-5}{x^2-9x+20}$, has a vertical asymptote at $x =$

(A) 0 only
(B) 4 only
(C) 5 only
(D) 0 and 4 only
(E) 0, 4, and 5

7. The volume of the region between two concentric spheres of radii 3 and 4 is

(A) 66
(B) 28
(C) 368
(D) 155
(E) 113

8. Under which conditions is $\dfrac{x-y}{xy}$ negative?

(A) $0 < y < x$
(B) $x < y < 0$
(C) $x < 0 < y$
(D) $y < x < 0$
(E) None of the above

9. If surface area of a sphere has the same numerical value as its volume, what is the length of the radius of this sphere?

(A) 1
(B) 2
(C) 3
(D) 4
(E) 6

10. If a and b are in the domain of a function f and $f(a) < f(b)$, which of the following must be true?

(A) $a = 0$ or $b = 0$
(B) $a < b$
(C) $a > b$
(D) $a \neq b$
(E) $a = b$

11. In $\triangle ABC$, $\sin A = \dfrac{\sqrt{2}}{2}$, $\sin B = \dfrac{\sqrt{3}}{3}$, and $BC = \sqrt{5}$ inches. The length of AC, in inches, is

(A) 3.0
(B) 3.9
(C) 3.5
(D) 1.8
(E) 4.0

12. $(x + y)(\dfrac{1}{x} + \dfrac{1}{y}) =$

(A) $\dfrac{x+y}{xy}$

(B) 1

(C) $\dfrac{2(x+y)}{xy}$

(D) $\dfrac{(x+y)^2}{xy}$

(E) $\dfrac{1}{(x+y)^2}$

13. What is the range of the data set 6, 8, 8, 13, 16, 16?

(A) 12
(B) 18
(C) 13
(D) 15
(E) 10

14. The length of the radius of the sphere $x^2 + y^2 + z^2 - 6y + 2z = 5$ is

(A) 3.16
(B) 3.38
(C) 3.87
(D) 3.74
(E) 3.46

15. Of the following lists of numbers,
    which has the biggest standard deviation?

    (A) 2, 5, 8
    (B) 3, 5, 9
    (C) 4, 6, 8
    (D) 1, 9, 18
    (E) 2, 8, 9

16. Which ordered number pair represents
    the center of the ellipse of $3x^2 + 4y^2 - 6x - 8y = 0$?

    (A) $(1, 3)$
    (B) $(2, 2)$
    (C) $(3, -1)$
    (D) $(1, 1)$
    (E) $(1, 5)$

17. If $m < 0$, the amplitude value of $2m \cdot \sin 2x$ is

    (A) 2
    (B) m
    (C) $2m$
    (D) $4m$
    (E) $|2m|$

18. If $f(x) = \log_2 x^a$ and $f(2) = 2.1$, then the value of a is

    (A) 1.3
    (B) 2.1
    (C) 0.3
    (D) 13.2
    (E) 32.5

19. Find the value of the remainder obtained when
    $x^4 - 4x^2 - x + 6$ is divided by $x + 2$.

    (A) 2
    (B) 4
    (C) 6
    (D) 8
    (E) 10

20. $\lim\limits_{x \to \infty} \left(\dfrac{3x^2 + 5x - 6}{8x^2 + 3x + 1} \right) =$

    (A) $\dfrac{3}{8}$
    (B) 1
    (C) -5
    (D) $\dfrac{1}{5}$
    (E) This expression is undefined.

21. If x and y are real numbers, which is a function of x?

   (A) $x = 2y^4 - 3$
   (B) $y = 2x^4 + 1$
   (C) $y = \pm\sqrt{(4 - x^2)}$
   (D) $y < x^2 - 1$
   (E) $x = \cos y$

USE THIS SPACE FOR SCRATCH WORK

22. The circle $x^2 + y^2 = 16$ and the hyperbola $\dfrac{x^2}{4} - \dfrac{y^2}{9} = 1$

   intersect at points where the y-coordinate is

   (A) ± 1.41
   (B) ± 2.24
   (C) ± 10.00
   (D) ± 2.45
   (E) ± 2.88

23. If $\sin x = \cot x$,
   which of the following is a possible radian value of x?

   (A) -1.00
   (B) -0.52
   (C) 0.00
   (D) 0.52
   (E) 0.90

24. If $f(x) = 2x$ and $g(x) = x^2 + 1$,
   which of the following must be true?

   I. $f(x)g(x)$ is an odd function.
   II. $f(g(x))$ is an even function.
   III. $g(f(x))$ is an even function.

   (A) only I
   (B) only II
   (C) only III
   (D) only II and III
   (E) I, II, and III

25. If $f(x) = 2x + 1$ and $f(g(1)) = 7$,
   which of the following could be $g(x)$?

   (A) $7x - 4$
   (B) $5x + 7$
   (C) $5x - 7$
   (D) $5x + 3$
   (E) $-5x + 3$

201

26. If A is the angle formed by the line $4y = x + 5$ and the x-axis, then $\angle A$ equals

(A) 72°
(B) 56°
(C) 14°
(D) 0°
(E) -45°

27. The radius of a sphere is equal to the radius of the base of the cone. The height of a cone is equal to the radius of its base. The ratio of the volume of the *cone* to the volume of the *sphere* is

(A) $\dfrac{1}{3}$

(B) $\dfrac{1}{4}$

(C) $\dfrac{1}{12}$

(D) $\dfrac{1}{1}$

(E) $\dfrac{4}{3}$

28. If $\log_2 x = y$ and $\log_e 2 = a$, then

(A) $\log_e x = ay$

(B) $\log_e x = \dfrac{y}{a}$

(C) $\log_e y = \dfrac{x}{a}$

(D) $\log_e y = ax$

(E) none of these

29. An indirect proof of the statement "If $x < 0$, then $\sqrt{x}$ is not a real number" could begin with the assumption that

(A) $x = 0$
(B) $x > 0$
(C) $\sqrt{x}$ is real number
(D) $\sqrt{x}$ is not a real number
(E) x is nonnegative

202

30. Find the product of an infinite number of terms:
$5^{\frac{1}{3}} \times 5^{\frac{2}{9}} \times 5^{\frac{4}{27}} \times 5^{\frac{8}{81}} \times \ldots$

(A) $5^{\frac{1}{2}}$

(B) $\sqrt[4]{5}$

(C) 5^2

(D) 5

(E) 1

USE THIS SPACE FOR SCRATCH WORK

x	$f(x)$
-2	0
-1	3
0	1
1	0

31. If f is a polynomial of degree 3, four of whose values are shown in the table above, then $f(x)$ could equal

(A) $(x + \frac{1}{2})(x + 1)(x + 2)$

(B) $(x + 1)(x - 2)(x - \frac{1}{2})$

(C) $(x + 1)(x - 2)(x - 1)$

(D) $(x + 2)(x - \frac{1}{2})(x - 1)$

(E) $(x + 2)(x + 1)(x - 2)$

32. Write an equation of lowest degree, with real coefficients, if two of its roots are -2 and $1 - i$.

(A) $x^3 + 2x^2 + 4 = 0$
(B) $x^3 - 2x^2 - 4 = 0$
(C) $x^3 - 2x + 4 = 0$
(D) $x^3 - 2x^2 + 4 = 0$
(E) none of these

33. If $\tan \theta = x$, then, for all θ in the interval $0 \le \theta \le \frac{\pi}{2}$, $\sin \theta \cdot \cos \theta =$

(A) $\frac{1}{1+x^2}$

(B) $\frac{x}{1+x^2}$

(C) $\frac{x^2}{1+x^2}$

(D) $\frac{1}{\sqrt{1+x^2}}$

(E) $\frac{x}{\sqrt{1+x^2}}$

34. If $f(x) = \dfrac{\sqrt{k-1}}{x}$ for all nonzero real numbers, for what value of k does $f(f(x)) = x$?

    (A) only 1
    (B) only 0
    (C) all real numbers
    (D) all real numbers greater than 1
    (E) no real numbers

35. The vertex angle of an isosceles triangle is 60°. The length of the base is 6 centimeters. What is the area of this triangle?

    (A) 17.4
    (B) 44.9
    (C) 20.2
    (D) 16.6
    (E) 15.6

36. A coin is tossed four times. Find the probability of the event represented by the composite statement $\sim p \cap q$ if

    p: exactly two heads show
    q: at least two heads show

    (A) $\dfrac{1}{2}$

    (B) $\dfrac{5}{16}$

    (C) $\dfrac{1}{6}$

    (D) $\dfrac{1}{8}$

    (E) $\dfrac{3}{4}$

37. If the perimeter of an isosceles triangle is 16 and the altitude to the base is 4, find the length of the altitude to one of the legs.

    (A) 4.8
    (B) 6
    (C) 9.6
    (D) 10
    (E) Cannot be found on the basis of the given data

38. Two spheres, of radius 9 and 4, are resting on a plane table top so that they touch each other at point Q. How high is the point Q from the plane table top?

(A) 4
(B) 5
(C) 5.54
(D) 6.22
(E) 7

39. $\cos(150° + x) - \cos(150° - x)$ equals

(A) $\sqrt{2} \sin x$
(B) $-\sin x$
(C) $\sqrt{3} \cos x$
(D) $\sqrt{2} \cos x$
(E) $\sqrt{3} \sin x$

40. Which of the following could be the coordinates of the center of a circle tangent to the x-axis and the line $x = 1$?

(A) $(-1, 0)$
(B) $(-1, 2)$
(C) $(0, 2)$
(D) $(2, -2)$
(E) $(3, 1)$

41. What is the probability of getting more than two heads when flipping three coins?

(A) $\dfrac{3}{4}$

(B) $\dfrac{1}{4}$

(C) $\dfrac{7}{16}$

(D) $\dfrac{1}{8}$

(E) $\dfrac{3}{16}$

42. Which of the following is a cube root of $-1 + \sqrt{3}i$?

(A) $+i$
(B) $\sqrt[3]{2} (\cos 140° + i \sin 140°)$
(C) $\cos 90° + i \sin 90°$
(D) $\sqrt[3]{2} (\cos 160° + i \sin 160°)$
(E) $\sqrt[3]{2} (\cos 60° + i \sin 60°)$

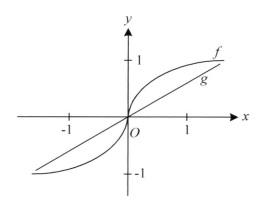

43. Portions of the graphs of f and g are shown above. Which of the following could be a portion of the graph of $f \cdot g$?

(A)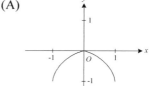

(B)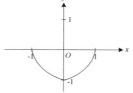

(C)

(D)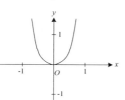

(E)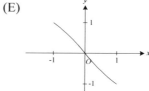

44. If the probability that Thomas will win the American Chess Championship is p and if the probability that James will win the World Chess Championship is q, what is the probability that only one of Thomas or James will win its respective championship?

(A) pq

(B) $p + q - 2pq$

(C) $|p - q|$

(D) $1 - pq$

(E) $2pq - p - q$

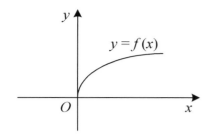

45. If the graph above shows the function $f(x)$, then which of the following could represent the inverse of $y = f(-x)$?

(A)
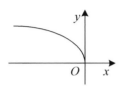

(B)
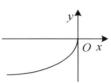

(C)
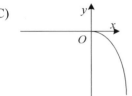

(D)
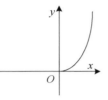

(E)
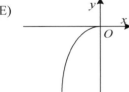

46. If $\sin x = \dfrac{3}{5}$ and $\dfrac{\pi}{2} \le x \le \pi$, then $\tan 2x =$

(A) $-\dfrac{7}{24}$

(B) $-\dfrac{24}{25}$

(C) $-\dfrac{3}{4}$

(D) $-\dfrac{24}{7}$

(E) $\dfrac{7}{25}$

47. Three consecutive terms, in order, of an geometric sequence are $x - \sqrt{5}$, $2x + \sqrt{2}$, and $4x - \sqrt{5}$. Then x equals

(A) 0.46
(B) 1.56
(C) 0.18
(D) 1.45
(E) 0.24

USE THIS SPACE FOR SCRATCH WORK

48. If $f(x) = (x - a)(x - b)$, how must a and b be related so that the graph of $f(x + 3)$ will be symmetric about the y-axis?

(A) $b = 6 + a$
(B) $b = 0$, a is any real number
(C) $b = 3a$
(D) $b = 6a$
(E) $b = 6 - a$

49. The area of the region enclosed by the graph of the polar curve $r = \dfrac{-2}{\sin\theta - 2\cos\theta}$ and the x- and y-axes is

(A) 0.48
(B) 0.50
(C) 0.52
(D) 0.98
(E) 1.00

50. What is the vertex of $y = f(x) = x^2 - 3x + 2$?

(A) $(1, 2)$
(B) $(2, 3)$
(C) $(4, 2)$
(D) $(-3, 1)$
(E) $(1.5, -0.25)$

Answer Key
Model Test No. 06

1	D	14	C	27	B	40	B
2	C	15	D	28	A	41	D
3	B	16	D	29	C	42	D
4	A	17	E	30	D	43	D
5	E	18	B	31	D	44	B
6	B	19	D	32	C	45	C
7	D	20	A	33	B	46	D
8	B	21	B	34	D	47	C
9	C	22	E	35	E	48	E
10	D	23	E	36	B	49	E
11	D	24	E	37	A	50	E
12	D	25	A	38	C		
13	E	26	C	39	B		

How to Score the SAT Subject Test in Mathematics Level 2

When you take an actual SAT Subject Test in Mathematics Level 2, your answer sheet will be "read" by a scanning machine that will record your responses to each question. Then a computer will compare your answers with the correct answers and produce your raw score. You get one point for each correct answer. For each wrong answer, you lose one-fourth of a point. Questions you omit (and any for which you mark more than one answer) are not counted. This raw score is converted to a scaled score that is reported to you and to the colleges you specify.

Finding Your Raw Test Score

STEP 1: Table A lists the correct answers for all the questions on the Subject Test in Mathematics Level 2 that is reproduced in this book. It also serves as a worksheet for you to calculate your raw score.

- Compare your answers with those given in the table.
- Put a check in the column marked "Right" if your answer is correct.
- Put a check in the column marked "Wrong" if your answer is incorrect.
- Leave both columns blank if you omitted the question.

STEP 2: Count the number of right answers.
Enter the total here: _____

STEP 3: Count the number of wrong answers.
Enter the total here: _____

STEP 4: Multiply the number of wrong answers by .250.
Enter the product here: _____

STEP 5: Subtract the result obtained in Step 4 from the total you obtained in Step 2.
Enter the result here: _____

STEP 6: Round the number obtained in Step 5 to the nearest whole number.
Enter the result here: _____

The number you obtained in Step 6 is your raw score.

Scaled Score Conversion Table
Subject Test in Mathematics Level 2

Raw Score	Scaled Score	Raw Score	Scaled Score	Raw Score	Scaled Score
50	800	28	630	6	470
49	800	27	630	5	460
48	800	26	620	4	450
47	800	25	610	3	440
46	800	24	600	2	430
45	800	23	600	1	420
44	800	22	590	0	410
43	790	21	580	-1	400
42	780	20	580	-2	390
41	770	19	570	-3	370
40	760	18	560	-4	360
39	750	17	560	-5	350
38	740	16	550	-6	340
37	730	15	540	-7	340
36	710	14	530	-8	330
35	700	13	530	-9	330
34	690	12	520	-10	320
33	680	11	510	-11	310
32	670	10	500	-12	300
31	660	9	490		
30	650	8	480		
29	640	7	480		

1. If $8x + 12 = \dfrac{k}{3}(2x + 3)$ for all x, then $k =$

   (A) $\dfrac{1}{4}$

   (B) 3
   (C) 4
   (D) 12
   (E) 24

$8x + 12 = \dfrac{k}{3}(2x + 3)$

$\therefore 3(8x + 12) = k(2x + 3)$, or $24x + 36 = 2kx + 3k$
$\therefore 2k = 24$ $\therefore k = 12$
Or, we may solve in other way,

$8x + 12 = 4(2x + 3) = \dfrac{k}{3}(2x + 3)$

$\therefore 4 = \dfrac{k}{3}$ $\therefore k = 12$

Ans. (D)

2. Which of the following is an equation of a line perpendicular to $y = 3x - 2$?

   (A) $y = 2x + 3$ (B) $y = -3x + 2$

   (C) $y = -\dfrac{1}{3}x + 5$ (D) $y = \dfrac{1}{3}x - 2$

   (E) $y = \dfrac{1}{-3x + 2}$

$y = 3x - 2$
$\therefore$ Slope of this line is $m_1 = 3$.
To be perpendicular, we have $m_1 \cdot m_2 = -1$, $3m_2 = (-)1$
$\therefore$ The slope of the other line $m_2 = (-)\dfrac{1}{3}$

Ans. (C)

3. Find the solution set of the inequality $x^2 - 3x - 4 < 0$.

   (A) $x > -1$ (B) $-1 < x < 4$
   (C) $x > 4$ and $x < -1$ (D) $x > 4$ or $x < -1$
   (E) $x < 4$

$x^2 - 3x - 4 < 0$, or $(x - 4)(x + 1) < 0$
$\therefore -1 < x < 4$.

Ans. (B)

4. If $f(x, y) = (\ln \sqrt{2}x^3)e^{2\sqrt{y}}$, what is the approximate value of $f(\sqrt{2}, 2)$?

   (A) 23.45   (B) 24.35
   (C) 25.34   (D) 25.43
   (E) 27.25

Replacing $x = \sqrt{2}$, and $y = 2$,
we get $f(\sqrt{2}, 2) = (\ln \sqrt{2} \cdot (\sqrt{2})^3)e^{2\sqrt{2}} = \ln (\sqrt{2})^4 \cdot e^{2\sqrt{2}}$
$= (\ln 2^2) e^{2\sqrt{2}} = (2 \cdot \ln 2) e^{2\sqrt{2}} = 23.45$

Ans. (A)

5. $\dfrac{\sin\frac{2\pi}{3}\cos\frac{\pi}{3}}{\tan 45°} =$

   (A) $\dfrac{\sqrt{3}}{2}$ (B) $-\dfrac{\sqrt{3}}{4}$ (C) $\dfrac{\sqrt{6}}{4}$ (D) $-\dfrac{\sqrt{3}}{2}$ (E) $\dfrac{\sqrt{3}}{4}$

Using calculator in radian mode, we get $45° = \dfrac{\pi}{4}$.

$\therefore \dfrac{\sin\frac{2\pi}{3}\cos\frac{\pi}{3}}{\tan\frac{\pi}{4}} = \dfrac{\frac{\sqrt{3}}{2} \times \frac{1}{2}}{1} = \dfrac{\sqrt{3}}{4}$

Ans. (E)

6. The graph of the rational function f, where $f(x) = \dfrac{x-5}{x^2-9x+20}$, has a vertical asymptote at $x =$

   (A) 0 only   (B) 4 only
   (C) 5 only   (D) 0 and 4 only
   (E) 0, 4, and 5

$f(x) = \dfrac{x-5}{(x-4)(x-5)} = \dfrac{1}{(x-4)}$, when $x \neq 5$.

Here, the removable discontinuity occurs at $x = 5$, and $x = 4$ serves as a vertical asymptote.

Ans. (B)

7. The volume of the region between two concentric spheres of radii 3 and 4 is

   (A) 66   (B) 28   (C) 368   (D) 155   (E) 113

$$V = \frac{4}{3}\pi(R^3 - r^3) = \frac{4}{3}\pi(4^3 - 3^3) = 154.98$$

Ans. (D)

8. Under which conditions is $\frac{x-y}{xy}$ negative?

   (A) $0 < y < x$
   (B) $x < y < 0$
   (C) $x < 0 < y$
   (D) $y < x < 0$
   (E) None of the above

Let's just try "Back substitution" for our answer choices.
Then, for (A), let's try $x = 2$, $y = 1 \rightarrow \frac{2-1}{2(1)} = \frac{1}{2} > 0$ ∴ No
For (B), let's try $x = -2$, $y = -1$,
then $\frac{(-2)-(-1)}{(-2)(-1)} = -\frac{1}{2} < 0$ ∴ Yes

Ans. (B)

9. If surface area of a sphere has the same numerical value as its volume, what is the length of the radius of this sphere?

   (A) 1
   (B) 2
   (C) 3
   (D) 4
   (E) 6

The formula for the surface area of sphere is $S = 4\pi r^2$,
and the volume $V = \frac{4}{3}\pi r^3$.
∴ $S = V \rightarrow 4\pi r^2 = \frac{4}{3}\pi r^3$, or $1 = \frac{1}{3}r$
∴ $r = 3$

Ans. (C)

10. If a and b are in the domain of a function f and $f(a) < f(b)$, which of the following must be true?

   (A) $a = 0$ or $b = 0$
   (B) $a < b$
   (C) $a > b$
   (D) $a \neq b$
   (E) $a = b$

We may have different cases for $f(a) < f(b)$
(case 1)
∴ If $a < b$,
then $f(a) < f(b)$.

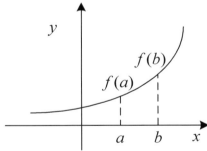

(case 2)
∴ If $a > b$,
then $f(a) < f(b)$.

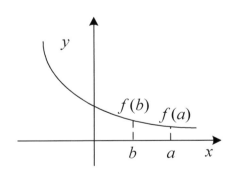

(case 3) If $a = b$ or $a = b = 0$, then $f(a) = f(b)$.
Therefore, to satisfy $f(a) < f(b)$ every time, the only answer is when $a \neq b$.

Ans. (D)

11. In $\triangle ABC$, $\sin A = \dfrac{\sqrt{2}}{2}$, $\sin B = \dfrac{\sqrt{3}}{3}$, and $BC = \sqrt{5}$ inches.
The length of AC, in inches, is

(A) 3.0
(B) 3.9
(C) 3.5
(D) 1.8
(E) 4.0

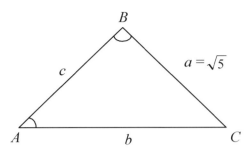

Referring to our previous note on the law of sine,

$\dfrac{\sin A}{a} = \dfrac{\sin B}{b} = \dfrac{\sin C}{c}$, We get $\dfrac{\sin A}{\sqrt{5}} = \dfrac{\sin B}{b}$, or $\dfrac{\frac{\sqrt{2}}{2}}{\sqrt{5}} = \dfrac{\frac{\sqrt{3}}{3}}{b}$.

$\therefore b\dfrac{\sqrt{2}}{2} = \sqrt{5}\dfrac{\sqrt{3}}{3}$

$\therefore b = \dfrac{\sqrt{15}}{3} \div \dfrac{\sqrt{2}}{2} = 1.825$

Ans. (D)

12. $(x + y)(\dfrac{1}{x} + \dfrac{1}{y}) =$

(A) $\dfrac{x+y}{xy}$ (B) 1 (C) $\dfrac{2(x+y)}{xy}$

(D) $\dfrac{(x+y)^2}{xy}$ (E) $\dfrac{1}{(x+y)^2}$

$(x + y)(\dfrac{1}{x} + \dfrac{1}{y}) = (x + y)\left(\dfrac{y+x}{xy}\right) = (x + y)\left(\dfrac{x+y}{xy}\right) = \dfrac{(x+y)^2}{xy}$

Ans. (D)

13. What is the range of the data set 6, 8, 8, 13, 16, 16?

(A) 12 (B) 18 (C) 13 (D) 15 (E) 10

The set {6, 8, 8, 13, 16, 16} runs from 6 to 16.
$\therefore$ The range is 16 - 6 = 10.

Ans. (E)

14. The length of the radius of the sphere
$x^2 + y^2 + z^2 - 6y + 2z = 5$ is

(A) 3.16
(B) 3.38
(C) 3.87
(D) 3.74
(E) 3.46

$x^2 + y^2 + z^2 - 6y + 2z = 5$ becomes $x^2 + (y^2 - 6y) + (z^2 + 2z) = 5$
Now, modifying this eq. into the complete square form,
we get $x^2 + (y - 3)^2 - 9 + (z + 1)^2 - 1 = 5$,
or $x^2 + (y - 3)^2 + (z + 1)^2 = 15$
$\therefore r = \sqrt{15} = 3.87$

Ans. (C)

15. Of the following lists of numbers,
which has the biggest standard deviation?

(A) 2, 5, 8
(B) 3, 5, 9
(C) 4, 6, 8
(D) 1, 9, 18
(E) 2, 8, 9

In statistics, the standard deviation means
"how far each point is away from the mean",

or the formula becomes, $\sigma = \sqrt{\dfrac{\sum_{i=1}^{n}(x_i - \bar{x})^2}{n}}$

But for this problem, without even using the above formula,
we intuitively know that the answer choice (D) 1, 9, 18
has the biggest difference between them.
Therefore, (D) has the biggest standard deviation.

Ans. (D)

16. Which ordered number pair represents the center of the ellipse of $3x^2 + 4y^2 - 6x - 8y = 0$?

    (A) (1, 3) (B) (2, 2) (C) (3, -1) (D) (1, 1) (E) (1, 5)

$3x^2 - 6x + 4y^2 - 8y = 3(x - 1)^2 + 4(y - 1)^2 = 7$
Therefore, the center is (1, 1)

Ans. (D)

17. If $m < 0$, the amplitude value of $2m \cdot \sin 2x$ is

    (A) 2
    (B) m
    (C) $2m$
    (D) $4m$
    (E) $|2m|$

$y = 2m \cdot \sin 2x$ has amplitude of $|2m|$,
and its period, $p = \dfrac{2\pi}{2} = \pi$.

Ans. (E)

(Note: The amplitude must have absolute value!!)

18. If $f(x) = \log_2 x^a$ and $f(2) = 2.1$, then the value of a is

    (A) 1.3   (B) 2.1   (C) 0.3   (D) 13.2   (E) 32.5

$f(x) = \log_2 x^a$
$\therefore f(2) = \log_2 (2)^a = a \cdot \log_2 (2) = a = 2.1$

Ans. (B)

19. Find the value of the remainder obtained when $x^4 - 4x^2 - x + 6$ is divided by $x + 2$.

    (A) 2   (B) 4   (C) 6   (D) 8   (E) 10

Let $P(x) = x^4 - 4x^2 - x + 6$. Then, as we refer to the "Remainder Theorem", the remainder becomes $P(-2)$.
$\therefore P(-2) = (-2)^4 - 4(-2)^2 - (-2) + 6 = 8$

Ans. (D)

20. $\lim\limits_{x \to \infty} \left(\dfrac{3x^2 + 5x - 6}{8x^2 + 3x + 1} \right) =$

    (A) $\dfrac{3}{8}$ (B) 1   (C) -5   (D) $\dfrac{1}{5}$

    (E) This expression is undefined.

Referring to the note on "Limit", since we have degree of two for both numerator and demoninator, the limit value becomes the ratio of the coefficient, that is, $\dfrac{3}{8}$.

Ans. (A)

21. If x and y are real numbers, which is a function of x?

    (A) $x = 2y^4 - 3$
    (B) $y = 2x^4 + 1$
    (C) $y = \pm\sqrt{(4 - x^2)}$
    (D) $y < x^2 - 1$
    (E) $x = \cos y$

To be a function, "the vertical line testing" must show that the line has only one intercept. (B) $y = 2x^4 + 1$ is the only one with one intercept with a vertical line.

Ans. (B)

22. The circle $x^2 + y^2 = 16$ and the hyperbola $\dfrac{x^2}{4} - \dfrac{y^2}{9} = 1$ intersect at points where the y-coordinate is

    (A) ±1.41
    (B) ±2.24
    (C) ±10.00
    (D) ±2.45
    (E) ±2.88

$x^2 + y^2 = 16$ and $\dfrac{x^2}{4} - \dfrac{y^2}{9} = 1$
Now, let $x^2 = 16 - y^2$, then $\dfrac{16 - y^2}{4} - \dfrac{y^2}{9} = 1$,
or simplifying it, $\dfrac{9(16 - y^2) - 4y^2}{36} = 1$
$\therefore 144 - 9y^2 - 4y^2 = 36 \rightarrow 13y^2 = 108$
$\therefore y^2 = \dfrac{108}{13}$ or, $y = \pm\sqrt{\dfrac{108}{13}} = \pm 2.88$

Ans. (E)

23. If $\sin x = \cot x$,
which of the following is a possible radian value of x?

(A) -1.00
(B) -0.52
(C) 0.00
(D) 0.52
(E) 0.90

$\sin x = \cot x$. Now, let's use graphing utility,

by setting $y_1 = \sin x$ and $y_2 = \dfrac{1}{\tan x}$, in radian mode.

Then, we get approximately, $x = 0.90$

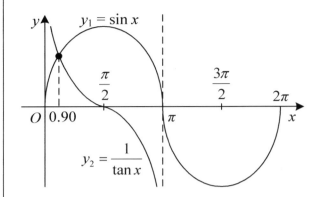

Ans. (E)

24. If $f(x) = 2x$ and $g(x) = x^2 + 1$,
which of the following must be true?

I. $f(x)g(x)$ is an odd function.
II. $f(g(x))$ is an even function.
III. $g(f(x))$ is an even function.

(A) only I
(B) only II
(C) only III
(D) only II and III
(E) I, II, and III

To be even function, it is symmetric to y-axis, and also to be odd function, it should be symmetric to origin $(0, 0)$.
Now, by trying graphing utility for (I), (II) and (III),

• $y_1 = f(x)g(x) = 2x(x^2 + 1)$ shows symmetric to $(0, 0)$
∴ True, odd function.

• (II) $y_2 = f(g(x)) = 2(x^2 + 1)$ shows symmetric to y-axis.
∴ True, even function.

• (III) $y_3 = g(5(x)) = (2x)^2 + 1$ shows symmetric to y-axis.
∴ True, even function.

Ans. (E)

25. If $f(x) = 2x + 1$ and $f(g(1)) = 7$,
which of the following could be $g(x)$?

(A) $7x - 4$
(B) $5x + 7$
(C) $5x - 7$
(D) $5x + 3$
(E) $-5x + 3$

$f(x) = 2x + 1, f(g(1)) = 7$
∴ $f(g(1)) = 2 \cdot g(1) + 1 = 7$
∴ $g(1) = 3$
Let $g(x) = 7x - 4$, then $g(1) = 7 - 4 = 3$

Ans. (A)

26. If A is the angle formed by the line $4y = x + 5$ and the x-axis, then $\angle A$ equals

(A) 72°
(B) 56°
(C) 14°
(D) 0°
(E) -45°

$4y = x + 5$, or $y = \dfrac{1}{4}x + \dfrac{5}{4}$

Now, the slope $m = \dfrac{1}{4}$, but slope m can be expressed as $\tan \theta$,
where θ is the angle formed by the line of the equation and x-axis.

∴ $m = \dfrac{1}{4} = \tan \theta$

∴ $\theta = \tan^{-1} \dfrac{1}{4} = 14°$

Ans. (C)

27. The radius of a sphere is equal to the radius of the base of the cone. The height of a cone is equal to the radius of its base. The ratio of the volume of the *cone* to the volume of the *sphere* is

(A) $\dfrac{1}{3}$

(B) $\dfrac{1}{4}$

(C) $\dfrac{1}{12}$

(D) $\dfrac{1}{1}$

(E) $\dfrac{4}{3}$

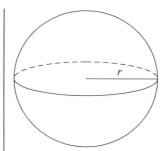 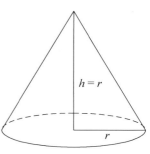

The volume of sphere is $V_1 = \dfrac{4}{3}\pi r^3$, and the volume of

cone is $V_2 = \dfrac{1}{3}\pi r^2 h = \dfrac{1}{3}\pi r^2(r) = \dfrac{1}{3}\pi r^3$

$\therefore \dfrac{V_2}{V_1} = \dfrac{\frac{1}{3}\pi r^3}{\frac{4}{3}\pi r^3} = \dfrac{1}{4}$

Ans. (B)

28. If $\log_2 x = y$ and $\log_e 2 = a$, then

(A) $\log_e x = ay$

(B) $\log_e x = \dfrac{y}{a}$

(C) $\log_e y = \dfrac{x}{a}$

(D) $\log_e y = ax$

(E) none of these

$\log_2 x = y$, $\log_e 2 = \ln 2 = a$

But, $\log_2 x = \dfrac{\ln x}{\ln 2} = y$

$\therefore \ln x = ay$

Ans. (A)

(Here, $\ln x = \log_e x$, where "ln" stands for "Natural log with base e".)

29. An indirect proof of the statement "If $x < 0$, then $\sqrt{x}$ is not a real number" could begin with the assumption that

(A) $x = 0$
(B) $x > 0$
(C) $\sqrt{x}$ is real number
(D) $\sqrt{x}$ is not a real number
(E) x is nonnegative

"An <u>indirect proof</u>" really means "Contra Positive" in logic, which means, the statement "if <u>P, then</u> Q", implies the contra positive statement of "if <u>not</u> Q, then <u>not</u> P". Applying this, we get, if $\sqrt{x}$ is a real number, then $x \geq 0$.

Ans. (C)

30. Find the product of an infinite number of terms:
$5^{\frac{1}{3}} \times 5^{\frac{2}{9}} \times 5^{\frac{4}{27}} \times 5^{\frac{8}{81}} \times \ldots$

(A) $5^{\frac{1}{2}}$

(B) $\sqrt[4]{5}$

(C) 5^2

(D) 5

(E) 1

$5^{\frac{1}{3}} \times 5^{\frac{2}{9}} \times 5^{\frac{4}{27}} \times 5^{\frac{8}{81}} \times \ldots = 5^{\frac{1}{3}+\frac{2}{9}+\frac{4}{27}+\frac{8}{81}+\cdots} =$

$5^{\frac{1}{3}(1+\frac{2}{3}+\frac{4}{9}+\frac{8}{27}+\cdots)} = 5^{\frac{1}{3}(1+\frac{2}{3}+(\frac{2}{3})^2+(\frac{2}{3})^3+\cdots)}$

Now, consider a geometric series:

$1 + \dfrac{2}{3} + \left(\dfrac{2}{3}\right)^2 + \left(\dfrac{2}{3}\right)^3 + \ldots$ which is an infinite series

with $a_1 = 1$, $r = \dfrac{2}{3}$

$\therefore$ The sum of the series, $S = \dfrac{1}{1-\frac{2}{3}} = 3$

$\therefore 5^{\frac{1}{3}(1+\frac{2}{3}+(\frac{2}{3})^2+(\frac{2}{3})^3+\cdots)} = 5^{\frac{1}{3}(3)} = 5^1 = 5$

Ans. (D)

x	$f(x)$
-2	0
-1	3
0	1
1	0

31. If f is a polynomial of degree 3, four of whose values are shown in the table above, then $f(x)$ could equal

(A) $(x + \frac{1}{2})(x + 1)(x + 2)$ (B) $(x + 1)(x - 2)(x - \frac{1}{2})$

(C) $(x + 1)(x - 2)(x - 1)$ (D) $(x + 2)(x - \frac{1}{2})(x - 1)$

(E) $(x + 2)(x + 1)(x - 2)$

By looking at the table, we get $f(-2) = 0$, $f(1) = 0$.
Now, referring to our previous note on "Factor Theorem",
$f(-2) = 0$ really means a factor of $(x + 2)$,
and $f(1) = 0$ also means a factor of $(x - 1)$.
Therefore, $f(x)$ must have factors of $(x + 2)$ and $(x - 1)$.
$\therefore f(x) = (x + 2)(x - 1) \cdot Q(x)$
From this, we need to find $f(x)$ that satisfies $f(0) = 1$,
as in the table. The only possible answer choice is,

(D) $f(x) = (x + 2)(x - \frac{1}{2})(x - 1)$, which has factors of $(x + 2)$ and $(x - 1)$, and also satisfies $f(0) = 1$.

Ans. (D)

32. Write an equation of lowest degree, with real coefficients,
if two of its roots are -2 and 1 - i.

(A) $x^3 + 2x^2 + 4 = 0$
(B) $x^3 - 2x^2 - 4 = 0$
(C) $x^3 - 2x + 4 = 0$
(D) $x^3 - 2x^2 + 4 = 0$
(E) none of these

One of the roots is (-2).
$\therefore (x - 2)$ factor.
The other complex root is $(1 - i)$, but the complex root must come with its conjugate, such as $(1 + i)$.
$\therefore f(x)$
$= (x + 2)[x - (1 - i)][x - (1 + i)]$
$= (x + 2)[(x - 1) + i][(x - 1) - i]$
$= (x + 2)[(x - 1)^2 - i^2]$
$= (x + 2)[x^2 - 2x + 1 - (-1)]$
$= (x + 2)(x^2 - 2x + 2)$
$= x^3 + 2x^2 - 2x^2 - 4x + 2x + 4$
$= x^3 - 2x + 4$

Ans. (C)

33. If $\tan \theta = x$, then, for all θ in the interval $0 \leq \theta \leq \frac{\pi}{2}$, $\sin \theta \cdot \cos \theta =$

(A) $\dfrac{1}{1+x^2}$

(B) $\dfrac{x}{1+x^2}$

(C) $\dfrac{x^2}{1+x^2}$

(D) $\dfrac{1}{\sqrt{1+x^2}}$

(E) $\dfrac{x}{\sqrt{1+x^2}}$

$\tan \theta = x$
$\therefore$

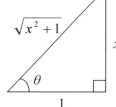

$\therefore \sin \theta = \dfrac{x}{\sqrt{x^2+1}}$, $\cos \theta = \dfrac{1}{\sqrt{x^2+1}}$

$\therefore \sin \theta \cdot \cos \theta = \dfrac{x}{\sqrt{x^2+1}} \cdot \dfrac{1}{\sqrt{x^2+1}} = \dfrac{x}{x^2+1}$

Ans. (B)

34. If $f(x) = \dfrac{\sqrt{k-1}}{x}$ for all nonzero real numbers, for what value of k does $f(f(x)) = x$?

(A) only 1
(B) only 0
(C) all real numbers
(D) all real numbers greater than 1
(E) no real numbers

$f(x) = \dfrac{\sqrt{k-1}}{x}$.

So, $f(f(x)) = f(\dfrac{\sqrt{k-1}}{x}) = \dfrac{\sqrt{k-1}}{\dfrac{\sqrt{k-1}}{x}} = x$.

This means that $\sqrt{k - 1}$ is for all real numbers, except for $k - 1 \neq 0$, and also for $k - 1 > 0$
$\therefore k > 1$

Ans. (D)

35. The vertex angle of an isosceles triangle is 60°. The length of the base is 6 centimeters. What is the area of this triangle?

(A) 17.4
(B) 44.9
(C) 20.2
(D) 16.6
(E) 15.6

Referring to the figure, we get the area of $\triangle ABC$

$$= \frac{1}{2} \text{(base)(height)}$$

$$= \frac{1}{2} (6)(3\sqrt{3})$$

$$= 9\sqrt{3} = 15.6$$

Ans. (E)

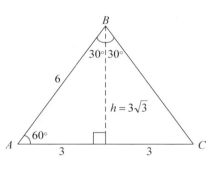

36. A coin is tossed four times. Find the probability of the event represented by the composite statement $\sim p \cap q$ if

p: exactly two heads show
q: at least two heads show

(A) $\frac{1}{2}$ (B) $\frac{5}{16}$ (C) $\frac{1}{6}$ (D) $\frac{1}{8}$ (E) $\frac{3}{4}$

Event $\sim p$ means the negation of P, that is, 0H, 1H, 3H or 4H. Also, event q means at least two heads, that is, 2H, 3H or 4H.
Event $(\sim p \cap q) = (0H, 1H, 3H, 4H) \cap (2H, 3H, 4H) = (3H, 4H)$.
That is, $_4C_3 + {_4C_4} = 4 + 1 = 5$.
But total number of possible events is $2 \times 2 \times 2 \times 2 = 16$.

$\therefore$ The probability becomes $\frac{5}{16}$

Ans. (B)

37. If the perimeter of an isosceles triangle is 16 and the altitude to the base is 4, find the length of the altitude to one of the legs.

(A) 4.8
(B) 6
(C) 9.6
(D) 10
(E) Cannot be found on the basis of the given data

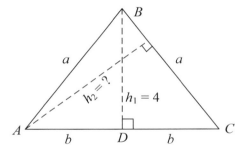

Here, instead of solving mathematically, let's try with our intuition.
That is, using the triangle $\triangle ABD$, we may try 3:4:5 ratio.

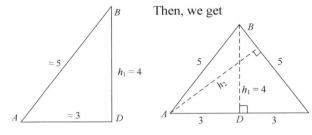

Therefore, the area of $\triangle ABC = \frac{1}{2}(6)(4) = 12$.

Now, the area of 12 also can be calculated by having $\overline{BC} = 5$ as the base and h_2 as its height.

$\therefore 12 = \frac{1}{2} \overline{BC} \cdot h_2 = \frac{1}{2}(5) \cdot h_2$

$\therefore h_2 = 4.8$

Ans. (A)

38. Two spheres, of radius 9 and 4, are resting on a plane table top so that they touch each other at point Q. How high is the point Q from the plane table top?

(A) 4
(B) 5
(C) 5.54
(D) 6.22
(E) 7

Given the information of the two spheres, we have the following figures:

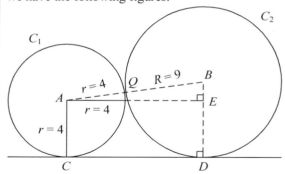

Here, $\overline{BD} = R = 9$, and $\overline{ED} = r = 4$
$\therefore \overline{BE} = 5$
Now, with the above information, we get

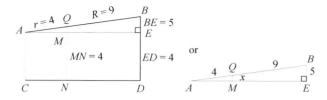

Using the ratio of $\triangle AQM$ and $\triangle ABE$ by similarity,

we get $\dfrac{\overline{AQ}}{\overline{AB}} = \dfrac{x}{5}$ or $\dfrac{4}{13} = \dfrac{x}{5}$

$\therefore x = \dfrac{20}{13} = 1.54$

$\therefore \overline{QN} = x + 4 = 1.54 + 4 = 5.54$

Ans. (C)

39. $\cos (150° + x) - \cos (150° - x)$ equals

(A) $\sqrt{2} \sin x$
(B) $-\sin x$
(C) $\sqrt{3} \cos x$
(D) $\sqrt{2} \cos x$
(E) $\sqrt{3} \sin x$

Refer to $\cos (A + B) = \cos A \cdot \cos B - \sin A \cdot \sin B$.
Then $\cos (150° + x) - \cos (150° - x)$
$= (\cos 150° \cdot \cos x - \sin 150° \cdot \sin x) -$
$(\cos 150° \cdot \cos x + \sin 150° \cdot \sin x)$

$= (-)2 \cdot \sin 150° \cdot \sin x = (-)2 \cdot \dfrac{1}{2} \sin x = -\sin x$

But, for this problem, I strongly recommend to use a calculator by letting $x = 30°$ or some other angle.
Then $\cos (150° + 30°) - \cos (150° - 30°)$

$= \cos 180° - \cos 120° = -1 - (-\dfrac{1}{2}) = -\dfrac{1}{2}$

Try answer choice (A) $\sqrt{2} \sin 30° = \sqrt{2} \dfrac{1}{2} = \dfrac{1}{\sqrt{2}}$ $\therefore$ No!

(B) $- \sin 30° = -\dfrac{1}{2}$ $\therefore$ Yes

Ans. (B)

40. Which of the following could be the coordinates of the center of a circle tangent to the *x*-axis and the line *x* = 1?

(A) (-1, 0)
(B) (-1, 2)
(C) (0, 2)
(D) (2, -2)
(E) (3, 1)

Given the information, we get

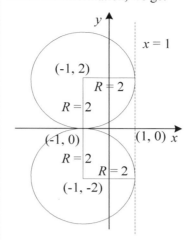

Ans. (B)

41. What is the probability of getting more than two heads when flipping three coins?

(A) $\dfrac{3}{4}$ (B) $\dfrac{1}{4}$ (C) $\dfrac{7}{16}$ (D) $\dfrac{1}{8}$ (E) $\dfrac{3}{16}$

Getting more than 2 heads will be only 3H out of tossing 3 coins.
$\therefore {}_3C_3 = 1$
But the total possible number of event will be $2 \times 2 \times 2 = 8$.
$\therefore p(3H) = \dfrac{1}{8}$

Ans. (D)

42. Which of the following is a cube root of $-1 + \sqrt{3}i$?

(A) $+i$

(B) $\sqrt[3]{2}\,(\cos 140° + i \sin 140°)$

(C) $\cos 90° + i \sin 90°$

(D) $\sqrt[3]{2}\,(\cos 160° + i \sin 160°)$

(E) $\sqrt[3]{2}\,(\cos 60° + i \sin 60°)$

Referring to our previous note on " Complex number" with polar form: $r = \sqrt{a^2 + b^2}$, $\theta = \tan^{-1} \dfrac{b}{a}$ for $z = a + bi$

$\therefore$ Here, $a = -1$, $b = \sqrt{3}$

$\therefore r = \sqrt{(-1)^2 + (\sqrt{3})^2} = \sqrt{4} = 2$,

$\theta = \tan^{-1} \dfrac{\sqrt{3}}{-1} = 120° + 360° \,(k)$.

Now, we get $z = -1 + \sqrt{3}i$
$= 2[\cos (120° + 360k) + i \sin (120° + 360k)]$
Therefore, by De Moivre's Theorem,

$z^{\frac{1}{3}} = 2^{\frac{1}{3}} \left(\cos \dfrac{120° + 360k}{3} + i \sin \dfrac{120° + 360k}{3}\right)$

When $k = 0$, $z^{\frac{1}{3}} = 2^{\frac{1}{3}}(\cos 40° + i \sin 40°)$

When $k = 1$, $z^{\frac{1}{3}} = 2^{\frac{1}{3}}(\cos 160° + i \sin 160°)$

When $k = 2$, $z^{\frac{1}{3}} = 2^{\frac{1}{3}}(\cos 280° + i \sin 280°)$

Ans. (D)

But for this problem, I strongly recommend to use calculator such as, (D) $[\sqrt[3]{2}\,(\cos 120° + i \sin 120°)]^3 = -1 + \sqrt{3}i$

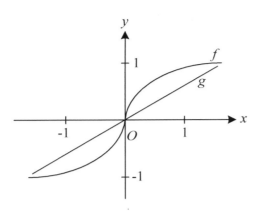

43. Portions of the graphs of f and g are shown above. Which of the following could be a portion of the graph of $f \cdot g$?

(A)
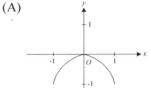

(B)
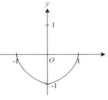

(C)

(D)
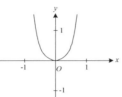

(E)
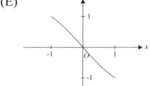

Since $f \cdot g$ on the left side of y-axis is positive, and approaches to zero as $x \to 0$ and $f \cdot g$ on the right side of y-axis is also positive, and grow bigger from zero, we get the graph of (D).

Ans. (D)

44. If the probability that Thomas will win the American Chess Championship is p and if the probability that James will win the World Chess Championship is q, what is the probability that only one of Thomas or James will win its respective championship?

(A) pq
(B) $p + q - 2pq$
(C) $|p - q|$
(D) $1 - pq$
(E) $2pq - p - q$

let p(Thomas win) $= p$, then p(Thomas lose) $= 1 - p$, and also let p(James win) $= q$, then p(James lose) $= 1 - q$.
Now, then the probability of only one of the two men will win be either;

• (case 1) If Thomas win, then James must lose. $\therefore p(1 - q)$ or
• (case 2) If James win, then Thomas must lose. $\therefore q(1 - p)$

Therefore, the probability becomes either case 1 or case 2.
$\therefore$ Prob $= p(1 - q) + q(1 - p) = p + q - 2pq$.

Ans. (B)

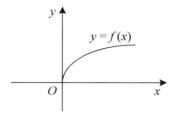

45. If the graph above shows the function $f(x)$, then which of
the following could represent the inverse of $y = f(-x)$?

(A)   (B)

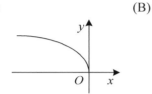

(C)

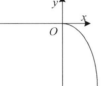

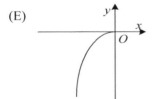

(D)   (E)

For the given function $y = f(x)$, $f(-x)$ is the function which is a reflection on y-axis.

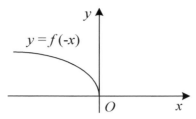

Now, this function must be symmetric to $y = x$ line to be inverse!
∴

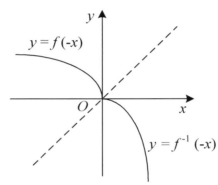

Ans. (C)

46. If $\sin x = \dfrac{3}{5}$ and $\dfrac{\pi}{2} \leq x \leq \pi$, then $\tan 2x =$

(A) $-\dfrac{7}{24}$

(B) $-\dfrac{24}{25}$

(C) $-\dfrac{3}{4}$

(D) $-\dfrac{24}{7}$

(E) $\dfrac{7}{25}$

$\sin x = \dfrac{3}{5}$

Now, from the triangle shown above, we get $\tan x = \dfrac{3}{4}$.

But our domain $\dfrac{\pi}{2} \leq x \leq \pi$ implies it is in the 2nd quadrant,

where $\tan x = -\dfrac{3}{4}$. Now, $\tan 2x = \dfrac{2\tan x}{1-\tan^2 x}$

$= \dfrac{2\left(-\dfrac{3}{4}\right)}{1-\left(-\dfrac{3}{4}\right)^2} = \dfrac{-\dfrac{6}{4}}{\dfrac{7}{16}} = \left(-\dfrac{6}{4}\right)\left(\dfrac{16}{7}\right) = -\dfrac{24}{7}$

Or, you may use the fact that $x = \sin^{-1}\dfrac{3}{5} = 36.9°$

∴ $\tan 2x = \tan 2(36.9°) = \tan 73.8° = 3.44$, but x is in the second quadrant, thus, the sign of tangent should be negative.

∴ The answer is $(-)3.44$ or $(-)\dfrac{24}{7}$

Ans. (D)

223

47. Three consecutive terms, in order, of an geometric sequence are $x - \sqrt{5}$, $2x + \sqrt{2}$, and $4x - \sqrt{5}$. Then x equals

(A) 0.46
(B) 1.56
(C) 0.18
(D) 1.45
(E) 0.24

In Geometric sequence, the square of the middle term is equal to the product of the neighboring terms. That is

$$\boxed{a_n^2 = a_{n-1} * a_{n+1}}$$

(eg) 2, 4, 8, 16, ...; $4^2 = 2 \times 8$, $8^2 = 4 \times 16$ ...

Therefore, $(2x + \sqrt{2})^2 = (x - \sqrt{5})(4x - \sqrt{5})$

$\therefore 4x^2 + 4\sqrt{2}x + 2 = 4x^2 - 4\sqrt{5}x - \sqrt{5}x + 5$

$\therefore 4\sqrt{2}x + 5\sqrt{5}x = 3$

$\therefore x(4\sqrt{2} + 5\sqrt{5}) = 3$

$\therefore x = \dfrac{3}{4\sqrt{2}+5\sqrt{5}} = 0.178$

Ans. (C)

48. If $f(x) = (x - a)(x - b)$, how must a and b be related so that the graph of $f(x + 3)$ will be symmetric about the y-axis?

(A) $b = 6 + a$
(B) $b = 0$, a is any real number
(C) $b = 3a$
(D) $b = 6a$
(E) $b = 6 - a$

$f(x) = (x - a)(x - b) = x^2 - (a + b)x + ab$

$\therefore f(x + 3) = (x + 3)^2 - (a + b)(x + 3) + ab$

$= x^2 + 6x + 9 - (a + b)x - 3(a + b) + ab$

Now, for $f(x + 3)$ to be symmetric to y-axis, we should not have x^1 term!!

$\therefore 6x - (a + b)x = 0$ or $(6 - a - b)x = 0$

$\therefore 6 - a - b = 0$

$\therefore b = 6 - a$

Ans. (E)

49. The area of the region enclosed by the graph of the polar curve $r = \dfrac{-2}{\sin\theta - 2\cos\theta}$ and the x- and y-axes is

(A) 0.48
(B) 0.50
(C) 0.52
(D) 0.98
(E) 1.00

$r = \dfrac{-2}{\sin\theta - 2\cos\theta} \rightarrow r(\sin\theta - 2\cos\theta)$

$= -2 \rightarrow r\sin\theta - 2r\cos\theta = -2$

But $x = r\cos\theta$ and $y = r\sin\theta$

$\therefore y - 2x = -2$, or $y = 2x - 2$

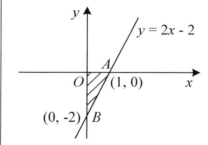

$\therefore$ area of $\triangle OAB = \dfrac{1}{2}(1)(2) = 1$

Ans. (E)

50. What is the vertex of $y = f(x) = x^2 - 3x + 2$?

(A) (1, 2)
(B) (2, 3)
(C) (4, 2)
(D) (-3, 1)
(E) (1.5, -0.25)

$y = f(x) = x^2 - 3x + 2$ has vertex at $x = -\dfrac{b}{2a}$.

$\therefore x = -\left(\dfrac{-3}{2}\right) = 1.5$

$\therefore f(1.5) = (1.5)^2 - 3(1.5) + 2 = 2.25 - 4.5 + 2 = -0.25$

$\therefore$ vertex, $v = (1.5, -0.25)$

Ans. (E)

44333348R00129

Made in the USA
Middletown, DE
03 June 2017